UNLOCKING YOUR BOWELS FOR BETTER HEALTH

Many in the healing arts believe that "Death begins in the colon!"

Chapter 1

* * *

The total volume of food, drink and gastro-intestinal secretions is about 2½ gallons per day. Only about 1/5th of a pint is finally lost through fecal elimination.

Chapter 2

* * *

Your stool can reflect a great deal about your eating and living habits and whether your digestive organs are functioning efficiently. A chart in this book on stool appearance is revealing!

Chapter 3

* * *

You can have one bowel movement a day and still be constipated! A normal stool should be almost odor free and not offensive.

Chapter 4

* * *

Crohn's Disease and Colitis are frustrating diseases that medically offer no satisfactory permanent relief.

Chapter 6

* * *

Excessive gas (flatulence) may be a symptom of a health problem such as gallstones, asthma or even heart disease.

Chapter 8

* * *

There are 11 known causes of diarrhea. In diarrhea you lose vitally important minerals and salts, especially potassium salts.

Chapter 10

* * *

Constipation is the cause of hiatal hernia. Most people do not know the natural position to have a bowel movement.

Chapter 12

* * *

Garlic, raw juices, bran and yogurt may prove highly beneficial to correcting your bowel problems!

Chapter 13

* * *

Medical doctors and nutritionists differ as to what is a normal, healthy transit time for a healthy bowel. Transit time means from the time you first eat a meal until the time the residue from the meal is eliminated.

Chapter 16

* * *

All this and much more (including Charts and Illustrations) **. . . you will find in the 17 chapters of this revealing book that shows how you can** *UNLOCK YOUR BOWELS FOR BETTER HEALTH!*

Important Legal Notice

The Medical Approach
versus
The Nutritional Approach
To
BOWEL PROBLEMS

by Salem Kirban

Published by Salem Kirban, Inc., Kent Road, Huntingdon Valley, Pennsylvania 19006. Copyright © 1981 by Salem Kirban. Printed in United States of America. All rights reserved, including the right to reproduce this book or portions thereof in any form.

Library of Congress Catalog Card No. 81-83270
ISBN 0-912582-41-3

ACKNOWLEDGMENTS

To **Estelle Bair Composition** for excellent craftsmanship in setting the type.

To **Walter W. Slotilock,** Chapel Hill Litho, for negatives.

To **Koechel Designs,** for an excellent cover design.

To **Bob Jackson,** for medical illustrations on pages 16, 20, 34, 38, 39, 42, 52, 79, 87, 105.

And special thanks to the following publishers for graciously making available medical illustrations for this book:

Intermed Communications, Inc., Horsham, Pennsylvania 19004
Illustrations reprinted with permission from Diseases, Copyright© 1981.

J.B. Lippincott Company, Philadelphia, Pennsylvania 19105
Illustrations reprinted with permission from Textbook of Medical-Surgical Nursing by L. Brunner and D. Suddarth, ed. 4, Copyright© 1980.

Mitchell Beazley Publishers, Ltd., England
Atlas of the Body and Mind, Copyright© Mitchell Beazley Publishers, Ltd. 1976. Published in U.S.A. by **Rand McNally & Company.**

American Cancer Society, New York, N.Y.

CONTENTS

Special Features Include

1

DEATH BEGINS IN THE COLON

The Formula

Many in the healing arts believe that:

Death begins in the colon!

An elderly doctor upon retiring from his practice at the age of 93 . . . was asked what was his formula for long life. His answer:

Trust God . . .
and keep your bowels open!

The colon is considered the sewage system of the body. By abuse and neglect of proper nutrition, it can become a foul cesspool. Everything starts to back up. Suddenly one discovers he has a disease or several diseases. It is estimated that there are 36 different poisons that come from the colon. One can see how all kinds of degenerative diseases can result from a clogged colon.

COLON SEAT OF MANY ILLNESSES

Digestive
Problems

Because your intestinal tract begins at your lips and ends at the anus there are many digestive problems that can occur when your colon is not functioning properly. Such illnesses include:

Hiatal Hernia
Peptic Ulcer
Gastric Ulcer
Gastritis
Stomach Cancer
Pancreatitis
Cancer of the Pancreas
Hepatitis
Cirrhosis of the Liver
Cancer of Stomach or Colon
Gallstones
Crohn's Disease
Colitis
Diverticulitis
Appendicitis
Diarrhea
Constipation
Hemorrhoids
Tumors

This list is by no means complete but does cover the most frequent illnesses. Digestive disorders account for the largest amount of time lost from work. They are the number one cause of disability absence. Digestive disorders also comprise one-sixth of all hospital admissions. Approximately 40% of all cancer originates in the gastrointestinal tract. Nutritionists would place this percentage even higher!

NAPOLEON LOST A BATTLE

Napoleon's Problem

It could be said with some degree of accuracy that Napoleon lost the battle of Waterloo because of two problems:

1. The British Army
2. Hemorrhoids

On June 18, 1815, Napoleon was near the small town of Waterloo in Belgium. From information Napoleon received he knew he should attack the Duke of Wellington's British troops early in the day because Wellington's reinforcements had been delayed.

But Napoleon was not physically able either to sit on his horse or to command his troops that day. His problem: diarrhea and hemorrhoids. His physician gave him sitz

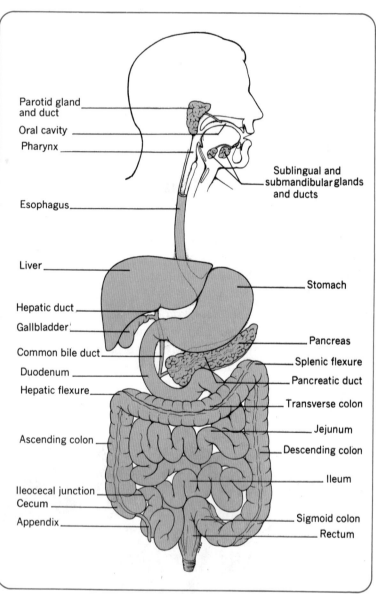

Parotid gland and duct

Oral cavity

Pharynx

Sublingual and submandibular glands and ducts

Esophagus

Liver

Stomach

Hepatic duct

Gallbladder

Pancreas

Common bile duct

Splenic flexure

Duodenum

Pancreatic duct

Hepatic flexure

Transverse colon

Jejunum

Ascending colon

Descending colon

Ileum

Ileocecal junction

Cecum

Appendix

Sigmoid colon

Rectum

Illustration courtesy J.B. Lippincott Company: Chaffee, RN, Greisheimer, E.: Basic Physiology and Anatomy, Philadelphia, J.B. Lippincott.

baths to alleviate the problem. Thus Napoleon was not able to move his army until the afternoon. By that time, Wellington's reinforcements had arrived and Napoleon lost the battle of Waterloo. The rest is history.

HOW OUR DIGESTION WORKS

**Fueled
By
Food**

If we are to understand the problems that can occur in our bowels . . . we must first be aware of our digestion system functions.

Man is fueled by food. The food you eat is processed through your body in a miraculous fashion. Actually, your digestive tract is about a 26-foot-long tube from your mouth to the source of elimination.

Food first enters your mouth and this is where digestion begins. As you chew your food it is mixed with saliva to ease the passage down the esophagus. Every day your salivary glands produce about three pints of saliva. This contains the starch-digesting enzyme ptyalin (amylase).[1]

When we decide to swallow, the ball of

[1]Proper chewing of your food is essential in the first step to healthy digestion. Most people eat too fast and do not chew their food sufficiently. It is a known fact that fast-food fare restaurants purposely play fast music to encourage their customers to eat their food quickly so other customers can be seated. The chairs are also designed in such a way that they are at first comfortable . . . but after 10 minutes become very uncomfortable!

food passes into the esophagus. Here powerful waves of muscular contraction squeeze the food down into the stomach. This squeezing, propelling action is called *peristalsis*. Peristalsis action also occurs in your stomach and colon.

THE STOMACH

24-Hour Duty

The next stop for your food is your stomach. The stomach is like a gatekeeper on 24-hour duty. The principal role of the stomach is one of a storage tank where food can be kept until the small intestine is ready to receive it.

The adult stomach can hold up to two and a half pints. The gastric juice in the stomach starts the breakdown of the proteins in the food and kills any contaminating bacteria.

Gastric Juices

A normal stomach is an acid stomach. Even the sight or smell of food will stimulate the production of gastric juices. The three major components of gastric juice are: hydrochloric acid, digestive enzymes and mucus.

The digestive enzyme *pepsin* along with the hydrochloric acid starts the conversion of the various protein foods into the simpler amino acids. The amino acids are the building blocks of the body tissues. Another enzyme, *rennin*, plays an important role in the digestion of the milk protein.

The mucus protects the gastric lining from any risk of self-digestion.

In the complexities of digestion, the millions of tiny glands in the stomach enable the Vitamin B_{12} to be absorbed.

While the *ptyalin*-containing saliva that starts the digestion of starch . . . is alkaline; the gastric juice that starts the digestion of protein is normally acid. Starches require an alkaline medium for digestion so starch digestion ceases in the acidic stomach.

**Pity
The
Stomach**

The stomach has three layers of muscle of that enable it to squeeze, mix and knead its contents with unusual power. The end result is a thin soupy mixture called *chyme.*
The stomach, a J-shaped organ about 10 inches long, then starts to empty its contents into the small intestine. The initial emptying takes place about 15 minutes after you have eaten. A meal remains in the stomach, however, for as long as four hours. The bigger the meal, the more rapid the emptying.

Carbohydrate-rich food leaves the stomach first, followed by protein-rich food and finally fatty foods. Many nutritionists recommend eating your protein foods first so the hydrochloric acid can be fully utilized. They suggest eating your salad after your protein foods. Some believe this procedure makes for easier digestion.

The outlet of the stomach is a purse-string valve called the *pylorus.* The word, *pylorus,* is Latin for "*gatekeeper.*"

2

THE SMALL INTESTINE . . . MIRACLE WORKER

THE SMALL BOWEL

**The
Small
Intestine**

The *pyloric sphincter* opens and closes to allow food into the upper ten inches of the small intestine. This area is called the *duodenum.*

The small intestine is often referred to as the small bowel and the large intestine as the large bowel or colon.

The duodenum is a horseshoe-shaped tube. In the duodenum, the hydrochloric acid in the food arriving from the stomach is neutralized by alkaline digestive juices. Some of these digestive juices come from the pancreas, some come from the liver. These intestinal enzymes complete the breakup of food elements and prepare them for absorption into the lymphatic system and the portal vein.

**The
Function
Of
Bile**

The fats cause the duodenum to release a hormone, *cholecystokinin.* This makes the gall bladder contract. As a result of this action, bile (a thick green alkaline solution) is poured out and mixes with the food. Bile is made up of salts and pigments made in the liver and stored in the gall bladder. Bile salts are necessary for the efficient digestion of fats. The bile emulsifies the fats so that they are more easily acted on by enzymes, known as *lipase.*

Bile also lowers the surface tension of the intestinal mash making it more wettable. Bile is the source of most of the brown color of normal stools.

The enzyme *trypsin* completes the digestion of protein. The enzyme *amylopsin* completes the digestion of the starches.

While the secretions of the stomach are of an acid nature . . . the secretions of the small intestine are alkaline.

**Three
Parts**

The small bowel, also known as the small intestine, is about 20 feet long and is coiled in the abdomen. It is divided into three parts:

> Duodenum
> Jejunum
> Ileum

The small intestine presents one of the many mysteries of nutrition. It is here that most of the nutrients in the food are absorbed by the body. It is called the small intestine not because of its length, but because of its diameter. It is longer than the large intestine (colon) but the large intestine is much wider.

THE ABSORPTION PROCESS

**The
Villi**

The *villi*, fingerlike tentacles that line the inner walls, look like the pile of a carpet. And it is these villi which are the instruments of the absorption process.

The SMALL INTESTINE

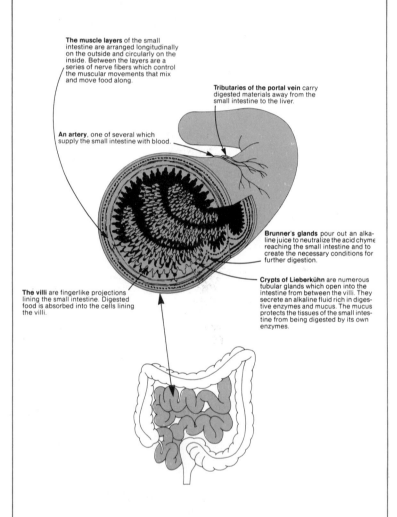

The muscle layers of the small intestine are arranged longitudinally on the outside and circularly on the inside. Between the layers are a series of nerve fibers which control the muscular movements that mix and move food along.

Tributaries of the portal vein carry digested materials away from the small intestine to the liver.

An artery, one of several which supply the small intestine with blood.

Brunner's glands pour out an alkaline juice to neutralize the acid chyme reaching the small intestine and to create the necessary conditions for further digestion.

Crypts of Lieberkühn are numerous tubular glands which open into the intestine from between the villi. They secrete an alkaline fluid rich in digestive enzymes and mucus. The mucus protects the tissues of the small intestine from being digested by its own enzymes.

The villi are fingerlike projections lining the small intestine. Digested food is absorbed into the cells lining the villi.

Top drawing shows cross section of the small intestine. The villi are fingerlike tentacles that line the inner walls. They look like the pile of a carpet and are vital to proper food absorption.

These sensitive villi are about 1/25th of an inch long. And a square inch of intestine holds about 3500 of these densely packed villi.

**Villi
Can Become
Villains**

The villi are your friends as long as you treat them right. But start loading them up with too many:

1. Starches and sugars
2. Fat, grease and fried foods
3. Foods loaded with additives

and they will let you know about it in no uncertain terms. And the villi will become villains because of your mistreatment!

This may be in the form of colicky, sharp abdominal pains, distention of the abdomen, nausea, vomiting, a change in appearance of the stool, loss of weight, blood in the stool.

The Name

The medical doctor may call your problem:

1. Mesenteric Vascular Occlusion
2. Intussusception
3. Sprue Syndrome
4. Chronic Ulcerative Colitis
5. Diverticulosis
6. Polyps

But whatever big name they use to call it, it could be as a result of your bad eating habits.

If all the villi were flattened out, the lining would give a total surface area of 350 square yards . . . about the size of a tennis court. This enormous surface makes the absorption of food highly efficient. But it also increases the area of cells exposed to wear and tear through improper nutrition. As much as four ounces of cells may be shed every day from your intestinal wall. This represents a potential loss of more than one ounce of protein alone!

Serves Two Functions

The villi act as a distributor and separator. They separate the valuable particles of protein, sugar and fat from a number of waste ingredients such as cellulose. The lymphatic channels in the villi carry fat elements to the body and a network of blood vessels transport digested carbohydrates and proteins to the liver.

The total volume of food, drink and gastrointestinal secretions is about 2-1/2 gallons per day. Only about 1/5th of a pint is finally lost through fecal elimination.

ON GUARD DUTY

An Important Valve

Once absorption of the valuable nutrients takes place, a water mix containing mainly fibrous waste, indigestible cellulose and other unwanted products remains. This now passes through the *ileocecal valve.*

This valve is like a watchman. Its duty is to keep food material from reentering the small intestine. This valve is located just above the appendix area in the lower right part of your abdomen. This flap of tissue (the ileocecal valve) opens to allow the unwanted products to pass to the colon . . . and then closes.

3

THE LARGE INTESTINE . . . KEY TO GOOD HEALTH

THE LARGE BOWEL

The
Colon

The colon (large intestine/large bowel) is about 5 feet long. It has four basic functions:

1. Reabsorbs excessive water from stool.
2. Excretion of some poisons and waste products from the blood.
3. Fermentation of some food residues by bacteria.
4. Storage of waste products along with bacteria, intestinal gas, until time of elimination.

The colon is divided into three segments:

1. The ascending colon is the first part of the large intestine.
2. The transverse colon runs horizontally across the abdomen.
3. The descending segment of the colon runs vertically down the left side connecting to the anal canal.

Putrefying bacteria are found in the lower part of the colon (the descending colon).

The COLON

Cecum

Hepatic flexure

Ileum

Appendix

Ascending Colon

Splenic flexure

Transverse Colon

Rectum

Anus

Sigmoid Colon

Descending Colon

**The
Final
Process**

No digestive enzymes are secreted in the colon, but an alkaline fluid aids in the completion of the digestion begun in the small intestines. The contents of the colon are gradually dehydrated until they assume the consistency of normal feces or even become quite hard.

The large intestine can store considerable amounts of fecal matter (that which you eliminate through a bowel movement).

Billions of bacteria live in the large intestine. They help disintegrate the fibers and cellulose of vegetable foods. The protein residue undergoes certain putrefactive changes.[1] When one has a bowel movement, there is a continual loss of these bacteria into the feces.[2]

**Release
Of
Toxins**

When this putrefaction takes place, a number of toxic by-products are released. In a healthy individual these poisons will be counteracted by the action of other bacteria in the colon or rendered harmless by the liver.

If these toxic by-products are not promptly neutralized, they can be damaging to the body. And illness will begin!

---

[1]*Putrefaction* is the decomposition of animal matter, especially protein. It is associated with the bad odors and poisonous products caused by certain kinds of bacteria and fungi.

[2]*Feces* is the stool or bowel movement you excrete by way of the anus.

THE FINAL STOP

Gentle, wavelike contractions (peristalsis) ultimately moves the contents of the bowel halfway around the colon. Eventually, it is carried to the end portion of the colon (the lower 8-10 inches) into the rectum. At the end of the rectum is the anal canal. The anal canal is about 2 inches long and ends at the anus. The anus has both internal and external muscles which can control the movement of the bowels.

Stool Appearance Vital

In a healthy individual, the stool, when eliminated, should be well formed and brownish in color. The appearance of the stool or bowel movement can reflect on one's state of health. A chart appears in this book which offers some guidelines projected by those in the healing arts.

Feces (stool) smell most on a meat diet, less on a vegetable diet. Intestinal gases vary in their potency of odor. Intestinal gases are a mixture of swallowed air and of gases produced by the intestinal bacteria.

The stool can reflect a great deal about our eating and living habits and whether our digestive organs are functioning efficiently.

In a normal digestive process, the appearance of stools reflect what was eaten almost one whole day earlier.

4

WHAT IS NORMAL?

**Once
A Day
Or
Once
In 3 Days!**

Medical doctors as a rule believe that a normal bowel movement is one where:

> An individual has a bowel movement
> once a day or
>   once in three days . . .
> whatever is his regular pattern.

Many nutritionists would disagree. Max Warmbrand writes:

> *Those who move their bowels daily
> often assume that their bowels
> are functioning normally,
> but this is not necessarily the case . . .*

> *Many persons may have a daily bowel
> movement and yet suffer from
> constipation or sluggishness or delay.*

> *When the muscles of the colon
> are tired or flabby,
> they do not possess the power to
> propel the waste material onward,
> or they push it forward
> only very slowly and sluggishly.*[1]

---

[1]Max Warmbrand, The Encyclopedia of Health and Nutrition (New York: Pyramid Books) 1974, pp. 69, 70.

**Two
To
Three
Times A Day!**

Warmbrand and other nutritionists believe that in a healty individual, bowel movements should occur twice a day (or even after each meal). It is interesting to note that healthy babies move their bowels after each feeding.

Where there is only one daily bowel movement (or one every two or three days), some material may be retained in the colon. This retained fecal matter becomes dry and hard and can lead to impaction. Impaction is when the fecal matter is pressed so firmly together that it is almost immovable. Nutritionists sometimes use a series of colonics (irrigation of the colon) to soften and remove this poisonous debris.

With fecal material adhering to the colon wall, the entire digestive process is upset and such a condition encourages several types of illness.

Therefore, even though your bowels move daily, (in impaction) elimination takes place in the center of the canal while the old encrusted fecal material sticks to the walls of your colon.

A HEALTHY COLON IS VITAL TO LIFE!

**Disease
Begins
In The
Bowel!**

Many nutritionists believe that many degenerative diseases begin in the bowels. They include in this category diseases that in themselves are not in the small or large intestine. This would include:

Pancreatitis
Cancer of the Pancreas
Gallbladder Cancer
Gallstones
Liver disorders

**Result
Of
Clogged
Colon**

While these specific illnesses are not covered in this book to any large extent, some believe these illnesses are the end result of a clogged colon. The poisons begin to back up and eventually the disease is evidenced in another organ of the body such as the stomach, pancreas, liver or gallbladder. Therefore, this book deals not with the symtomatic influence of the diseases just mentioned, but, rather with the possible origins of these diseases in the small and large intestine (the colon).

1
STOOL IDENTIFICATION

When there are changes from your normal digestion process an indication of this may show up in your stool. (*Stool, feces, bowel movement* . . . all refer to the same thing.) The appearance of your stool can tell you a great deal about your digestive mechanism. It can be a life-saving early warning system directing you to see your doctor.

Babies / Stool

Babies	Stool
Newborn baby	First few stools of newborn baby are black. This is normal as newborn passes an ointmentlike substance called *meconium.*
Few days old baby	Stool turns from black to greenish brown and then greenish yellow. By end of first week, stool becomes golden yellow. Normal stool has no offensive smell
Breast-fed baby Formula-fed baby	Stools are usually looser in consistency. Average breast-fed infant has two to four stools a day. Normal stool is salve-like and golden yellow. Formula-fed baby averages one or two stools a day.

Adults / Stool

Adults	Stool
High Protein diet (mostly meat)	Dark stool.
Dairy-rich diets	Light colored stool.

Significant bowel changes may be an indication of cancer. Alternating diarrhea with constipation or a fullness in the rectum which remains after a bowel movement should be reported to your doctor. Transitory bowel changes are not cancer indications. Changes in environment such as vacationing, eating spicy foods or a green apple or an emotional upset can bring a change in bowel habits for a brief time . . . and these are termed transitory. It is when these patterns persist that they should be reported to your physician.

Adult Stool Appearance

Possible Analysis

Adult Stool Appearance	Possible Analysis
Bright cherry red blood in stool	Bright red blood originates from the anus or rectum. Possibly hemorrhoids or other rectal problem.
Dark red blood in stool	The longer blood is in the digestive system, the darker it becomes. The farther up is the bleeding point. May indicate ulcers.
Yellow or Orange stools (Clay color)	Insufficient bile is mixed with the intestinal contents. May be the first sign of jaundice, liver disease or phosphorus poisoning.
Slate gray blackish stools	May be caused by iron medications taken in the treatment of anemia.
Very dark olive blue or olive grey "smeary" stools with very offensive odor.	May indicate a diet too rich in both protein and fat and also excessive putrefaction.
Jet black stools, offensive odor.	May indicate severe bleeding high in the intestinal tract. Bleeding may be from an ulcer in the stomach or duodenum. It could also come from a tooth extraction or from swallowing blood from a nosebleed. It could also come from the use of drugs which contain bismuth, iron, tannin, manganese or charcoal.

Adult Stool Appearance

Possible Analysis

Adult Stool Appearance	Possible Analysis
Stools that float	Fat eaten in the diet has not been digested. May indicate certain diseases of the pancreas or disorders of metabolism.
Blood and mucus in the stools. 15-20 stools a day, watery.	Possibly ulcerative colitis.
Mucus in stool.	Amount should be noted. Mucus is present in both normal and abnormal conditions. Many women in middle life will have mucus in the stool when no real disease exists. May appear as streaks or blobs.

Form and consistency of stool: Stool should be normally soft and formed. A hard, nodular stool is an indication of constipation. Flattened or ribbonlike stool may indicate a rectal obstruction or spastic colitis. A frothy stool indicates a fermentative condition. A greasy-like stool is typical in jaundice. A dark red stool may come from eating beets.

Odor: A normal stool should be almost odor free and not offensive. Offensive odor may indicate jaundice, acute indigestion, enteritis, constipation or improper digestion of proteins.

5

The Medical Approach To
DISEASES OF THE SMALL BOWEL

Celiac Disease (Sprue)

**Common
In
Children**

Celiac disease is common in children. In adults it is called Sprue. It is a problem caused because the contents of the small bowel (small intestine) are inadequately absorbed into the body.

The characteristics of Celiac/Sprue disease include lesions in the small bowel lining. Patients are usually sensitive to gluten. Gluten is a protein found in wheat, rye, oats or barley. It is also found in beans, cabbage, turnips, dried peas and cucumbers.

Other symptoms of this disease include:

> Abdominal distention
> Muscle wasting
> Diarrhea
> Stools pale and bad odor
> Abnormally low blood calcium

Medically, Celiac disease or Sprue is an insidious disease of unknown origin. The symptoms appear slowly and may initially only be an increase in gas in the colon and a gradual loss of appetite. Then there is loss of weight with bulky bowel movements that are light colored.

Anemia May Develop

Anemia often develops, and there may be small capillary hemorrhages under the skin due to a deficiency of Vitamin K. Vitamin K is one of the key factors in the blood-clotting mechanism. This disease not only affects people in the tropics but also in non-tropical countries.

Tropical Sprue is active in the Caribbean area, the Indian subcontinent and South-east Asia. The symptoms are similar.

Treatment is designed to eliminate foods which seem to aggravate the condition. These foods are initially eliminated from one's diet:

> All cereal grains which contain gluten.
> Milk and milk products.

Added to the diet are:

> Vitamin A
> B-complex
> Folic Acid

Medical treatment also includes, in some cases, the administration of cortisone hormones.

Calcium and potassium are often recommended as additional dietary aids.

Crohn's Disease
(Regional Enteritis)

An Inflammatory Disease

Crohn's Disease is an inflammatory disease of the gastrointestinal tract. Medically, it is of unknown origin.

While this disease is usually an affliction of

early middle age, it does also attack young people between the ages of 20 to 30.

It is sometimes called *Terminal Ileitis* when it affects the lower end of the ileum. This does not mean the disease is fatal. The ileum is the lower three-fifths of the small intestines and is about 12 feet long.

Ulcers May Develop

The inflammation process extends through the entire wall of the ileum. It is often accompanied by ulceration, scarring and *fibrosis* (an abnormal formation of fibrous tissue.)[1]

Regional Enteritis *(inflammation of the intestine)* is sometimes called Crohn's Disease after Dr. Burrill B. Crohn who described these lesions in the 1940's.

Early symtoms include:

Mid-abdominal cramps
Loose, non-bloody stools
Mild fever
Loss of appetite
Weight loss

Complications Can Occur

Patients with Crohn's Disease tend to have a high incidence of kidney stones and gallstones. Complications may include arthritis, red and painful nodules on the legs usually associated with rheumatism and ulcerative colitis.

The inflammation appears to begin in the lymph vessels under the mucous mem-

---

[1]Ileum refers to a part of the intestine. Do not confuse this with Ilium which is one of the bones of each half of the pelvis.

brane of the intestine. Soon it involves all the layers of the intestinal wall. There is swelling, dilation and eventually development of fibrous scar tissue. The areas affected frequently become permanently thickened, hardened and inelastic.

It is not uncommon for multiple areas of the intestine to be involved. Generally, about one or two feet of the ileum area are affected. It is believed medically that the disease is closely related to physical and emotional tension including continuous worry, conflict or pressure.

Root Cause Subject of Disagreement

There is some disagreement between physicians and psychiatrists as to the root cause of Crohn's Disease. The physician treats it mainly as a physical ailment. The psychiatrist maintains that these patients quite often become victims of this disease because of their conscious or subconscious feelings of anger, hostility and resentment.

Bowel habits change drastically from once or twice a day to sometimes loose bowel movements of 3-7 a day. A high fever is usually accompanied with abdominal pain in the lower right quadrant of the abdomen. It is sometimes mistaken for appendicitis.

If complications occur it can include perforation of the bowel and *fistulas*. A fistula is an abnormal passage from a normal cavity to another cavity or surface. These fistulas may occur to the bladder or vagina and in other areas. There is also a greater susceptibility to colon or rectal cancer.

Crohn's Disease (or Regional Enteritis) can be a disabling disease.

Initial Treatment

Initially it is treated medically and nutritionally. The diet recommended by physicians includes:

1. High vitamin intake
2. High calorie diet
3. Low residue diet

Raw fruits and vegetables are excluded because of their high residue content.

Drugs are administered . . . from a medical perspective . . . for a long period of time of six months to life.

Drugs Given

Drugs given include:

1. Sulfonamides
2. Ampicillin
3. Tetracycline

Corticosteroid therapy is considered useful medically in acute stages. In some cases, the drug, _azathioprine_ (Imuran) is given.

Crohn's Disease is both frustrating to the patient and to the physician.

Surgery Possible

If complications develop, surgery is performed and a portion of the small intestine is removed. Because of a smaller bowel area, the absorption surfaces of the small intestine are naturally diminished. Additional therapy is recommended to make sure this "shorter bowel" adjusts so there is no folic acid and Vitamin B_{12} deficiency.

**Recurrence
Frequent**

An ileostomy or colostomy (an opening of the small intestine or the large bowel through the abdominal wall) is not performed in this type of surgery.

The recurrence of this problem after surgery is 50% in a 5-year period. Recurrence is less frequent in patients over 50.[1]

---

[1]Complete information on CROHN'S DISEASE can be found in the companion book, The Medical Approach Versus The Nutritional Approach to **COLITIS/CROHN'S DISEASE.** For a copy of this book, send $6 ($5 plus $1 postage) to: Salem Kirban, Inc., Kent Road, Huntingdon Valley, Pennsylvania 19006.

ILEOSTOMY

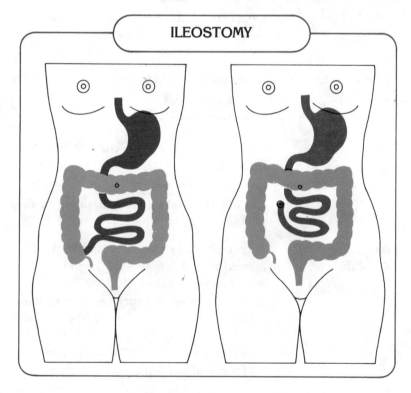

6

The Medical Approach To
DISEASES OF THE LARGE BOWEL
(Colon or Large Intestine)

Colitis

Inflammation Of Colon

There are various forms of Colitis. The term "*colitis*" should be applied only to actual inflammatory disease of the colon.

So-called "*spastic*" or "*mucous*" colitis is a functional disorder and is more properly described by the term "*irritable colon.*"

Chronic Ulcerative Colitis

Ulcerative Colitis is an inflammatory disease of the lining of the colon. The word "*chronic*" means that the disease is long and drawn out over a period of time. At times the entire colon may become inflamed. In some ways (rectal bleeding), it mimics Crohn's Disease. However Crohn's Disease affects the small intestine while Ulcerative Colitis affects the large intestine or colon.

The cause for both diseases is medically unknown. There is a risk that Ulcerative Colitis in its chronic state over a period of years can lead to cancer. It is known that populations with a high red-meat consumption have a high incidence of colon cancer.[1]

Doctors Disagree

The Harvard medical team states that:

*There is nothing to suggest that
lack of dietary fiber
causes these diseases
and fiber treatment
has no important advantage.
In fact, fiber can be hazardous
when bowel narrowing is present,
as is in the case in some patients
with Crohn's disease.*[2]

Frequent In Young Adults

Chronic Ulcerative Colitis is primarily a disease of adolescents and young adults but it may have its onset in any age group. Because of our poor nutritional lifestyle often the sins we commit in poor eating bring their toll in the form of Colitis in our 30's and 40's.

The disease is evidenced by multiple, irregular ulcerations in the colon. Repeated ulcerations lead to thickening of the wall of the colon with scar tissue developing. It has been found that many people with

[1]G. Timothy Johnson, M.D., Stephen E. Goldfinger, The Harvard Medical School Health Letter Book (Cambridge, Massachusetts: Harvard University Press), 1981, p. 298.

[2]Ibid., p. 30.

Chronic Ulcerative Colitis have a profile of being very sensitive and in some cases, neurotic (nervous).

Symptoms

Symptoms of Ulcerative Colitis include:

Bloody Diarrhea
(10 to 20 movements a day)
Lower abdominal cramps
Mild abdominal tenderness
Weight loss
Fever
Anemia
Loss of Appetite

Complete cure, medically, takes place in less than half the cases. Another 25% live on as intestinal cripples, never fully recovering but never quite sick enough to require surgical intervention. To save the lives and cure the remaining 25% of these patients, surgery is essential.[1]

**Surgery
Possible**

If surgery is required the ileum (a loop of the small intestine) is taken out into the abdominal wall and the bowel movements are channeled through this opening. This procedue is called an *ileostomy*. This allows the colon to heal. In about 10% of the cases, the ileostomy opening can be closed and the colon once more put into activity.

Basic treatment medically is bed rest and a high-protein, high caloric diet. Milk is banned.

---

[1]Robert E. Rothenberg, M.D., The Complete Surgical Guide (New York: Weathervane Books) 1974, p. 222.

SIGMOIDOSCOPY

A sigmoidoscope is an instrument about 10 inches long. It allows direct visualization of the entire rectum and the lower portion of the large bowel *(the sigmoid)*.

This hollow tube (the sigmoidoscope) is lighted. This enables the physician to note any inflammation or tumor growth. He will also be able to remove such polyps or burn them with an electric current. he can also use this instrument to take a biopsy.

Sigmoidoscopy is not a painful procedure. It is generally carried out in a doctor's office without anesthesia. The procedure may be slightly uncomfortable and embarrassing. It is best to have a cleansing enema an hour or two before the procedure. Otherwise, the physician will administer a Fleet-type enema about 15 minutes before he performs the sigmoidoscopy. Some doctors prefer that an enema not be used as the liquid residue may obscure proper vision.

After the sigmoidoscope is lubricated, it is introduced into the rectum, the physician depresses the bellows *(see illustration)* →
to introduce air into the rectum so it is fully expanded for examination. The procedure takes about 10 minutes.

 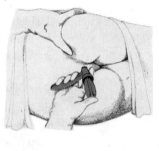

A **proctoscope** is an instrument for inspection of the rectum. In a proctoscope examination, the presence of blood, pus or mucus is observed as well as the mucous membrane. The proctoscope can identify a threadworm infection (commonly known as <u>pinworms</u>). Such an examination is also used to identify internal hemorrhoids.

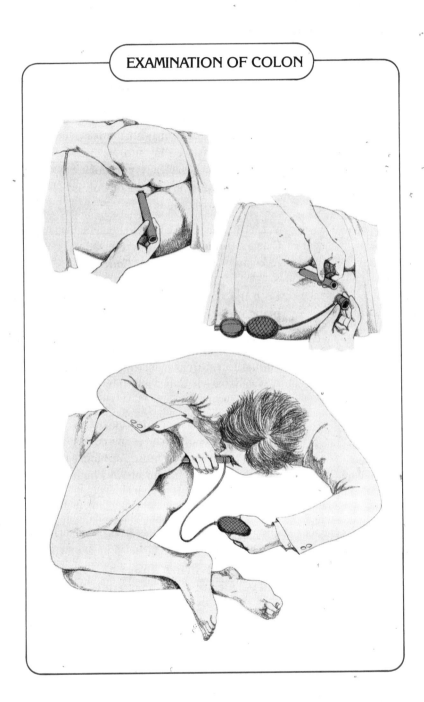

Drugs Used

Cortisone therapy and sulfonamides are the medical drug approach to the problem. Adrenocorticosteroid hormones such as Corticotropin (ACTH) and Hydrocortisone (Bio-Cortex, Cortef, Hydrocortone) are given during severe forms of Colitis. These naturally have side effects.[1]

The Irritable Bowel
(Mucous Colitis; Spastic Colitis)

A Familiar Disease

Both the names "mucous" and "spastic" are inaccurate names given to the Irritable Bowel disease. It is estimated that at one time or another some 50 to 75% of the population suffers from an Irritable Bowel. It is a disease brought about by our hurried pace of living.

Generally someone suffering from an Irritable Bowel is one who is overworked, has inadequate sleep, is tense, anxious, highly emotional and tends to always be hurried. Because of this they eat irregular and inadequate meals and abuse laxatives.

The Irritable Bowel Disease is not a serious disorder like Colitis.

Mismanaged living upsets the normally fine balance of functional elimination and

---

[1]Complete information on COLITIS can be found in the companion book, The Medical Approach Versus The Nutritional Approach To **COLITIS/CROHN'S DISEASE.** For a copy of this book, send $6 ($5 plus $1 postage) to: Salem Kirban, Inc., Kent Road, Huntingdon Valley, Pennsylvania 19006.

causes the colon to become irregular in its action.

Symptoms

Symptoms include:

    Constipation interspersed with Diarrhea
    Hard, dry stools covered with mucus
    Abdominal distress
    Gas (*flatus*)
    Loss of appetite in the morning
    Nausea
    Excessive belching
    Headaches
    Insomnia
    Excessive perspiration

The usual treatment medically is first to assure the patient that his disease is not serious. A mild sedative is sometimes given to help the patient eliminate stress conditions in his life. The patient is instructed to stay on a bland diet and avoid irritating foods and milk.

DIVERTICULOSIS and DIVERTICULITIS

Disease of Modern Society

If you are over 40, there is one chance in three that you have Diverticulosis.[1]

In Diverticu**losis** there is a weakness in the muscular wall of the bowel. This permits pouches of mucous membrane to sag through. Inflammation of these outpouchings is called Diverticu**litis.**

[1]Donald G. Cooley, Better Homes and Gardens After-40 Health & Medical Guide (Des Moines, Iowa: Meredith Corporation) 1980, p. 219.

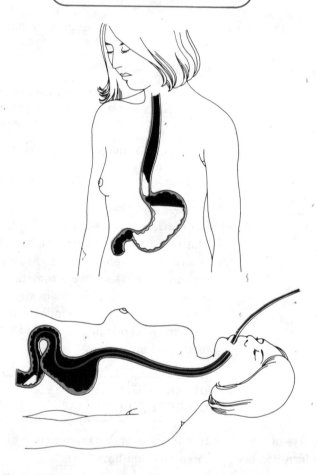

Top illustration: <u>Barium</u> meal used for an *"upper GI series."* Patient swallows barium sulfate which is opaque to X-rays. White area shows the barium which can reveal structures and possible abnormalities of the esophagus and stomach and even the duodenum.

Bottom illustration: <u>Endoscopy</u> . . . a fiber-optic instrument is introduced into the mouth. This flexible tube has a light at the end. Physician can thus photograph the esophagus, stomach and the small bowel.

DIGESTIVE TRACT TESTS

Several tests are used by medical doctors to determine digestive tract disorders.

Gastrointestinal (GI) series

Barium sulfate is used so that X-ray studies of the digestive tract can be made. Barium is opaque to X-rays. The barium is taken one of two ways:

1. Barium through the mouth

   For an upper GI series, the patient swallows a barium "meal." This is to study the esophagus, stomach and duodenum.

2. Barium enema (through the rectum)

   For a lower GI series, a barium enema is introduced into the rectum and colon. This is to study the colon and terminal small bowel. The barium's progress is noted with a fluoroscope. A motion picture may also be taken.

Cholecystography

This refers to X-ray studies of the gallbladder, usually for gallstones. The patient drinks a solution the night before. X-rays are taken the following morning. *Intravenous cholangiography* is where a dye is injected into the vein to visualize the bile ducts.

Endoscopy

A fiber-optic instrument is introduced into the mouth. The instrument is flexible. It enables the operator to visualize and photograph the esophagus, stomach, and the small bowel. He can also outline the liver (biliary) and pancreatic duct systems.

Colonoscope

The colonoscope is a narrow, flexible fiber-optic which is introduced into the rectum. It enables the operator to visualize the entire large bowel (colon) and even the end of the small bowel. It is also possible to remove polyps or tumors with the colonoscope.

Colonoscopy may be performed in the doctor's office on a simple examination couch or in a hospital operating room. An hour before the colonoscope, an intramuscular injection is given to cause a half-sleepy condition in the patient. A brief anesthesia may be given if procedure becomes uncomfortable. A day or two before the colonoscope bowel cleansing medication is given. On the day of the colonoscope an enema may be given.

This problem causes symptoms which some mistake for appendicitis. Pain is sometimes accompanied by nausea, vomiting, a slight increase in temperature and an elevation of white cell count. Often, however, the pain does not become as severe as appendicitis. An x-ray can identify Diverticulosis of the colon.

Diverticulosis only hits 5 or 10 per cent of young adults. It is more a disease of those in middle age and beyond. Chronic Diverticulosis is evidenced by bouts of alternating diarrhea and constipation, often with cramps and gas. If the bladder is affected, there is urinary frequency and urgency.

Complications Can Arise

Diagnosis is best made with a barium enema. Complications can arise in those persons who have Diverticulosis.

1. Perforation
   In severe cases, just like an inflamed appendix, an inflamed pouch on the colon can burst. This requires prompt surgery.

2. Bleeding
   If heavy bleeding occurs, surgery is also needed.

3. Abscesses and fistulas
   If the infection becomes blocked an abscess occurs. If the pathway of the colon is changed a fistula may occur. Surgical correction is often necessary.

4. Obstruction
   If the inflammation is acute, a narrowing of the colon could cause serious problems necessitating surgery.

Until recent years, medically, a diet <u>low</u> in residue was recommended. Now, however, most physicians agree that symptoms of Diverticular disease improve when a <u>high</u> residue (high fiber) diet is followed.

High Fiber Diet Beneficial

<u>Health Letter Books</u> reports:

> . . . there is excellent evidence
> *that symptoms of diverticular disease improve*
> *when high fiber diets are begun—*
> *though it may take as much as two months*
> *(and enduring a brief period of*
> *increased bloating and flatulence)*
> *before such benefits are realized.*[1]

Inflammation can occur when the pouches become plugged with fecal matter or undigested food. The pain that develops usually on the left side of the abdomen is sometimes called "left-sided appendicitis."

Acute Diverticulitis is treated with bed rest and some antibiotics. Sometimes sulfonamides are used.

Generally a high-fiver diet is recommended in the form of unprocessed bran and other forms of fiber.

Sometimes surgery is required in severe cases, occasionally with a colostomy.

---

[1]G. Timothy Johnson, M.D., Stephen E. Goldfinger, M.D., <u>The Harvard Medical School Health Letter Book</u> (Cambridge, Massachusetts: Harvard University Press) 1981, p. 29.

COLOSTOMY

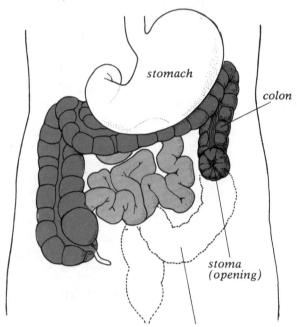

stomach

colon

stoma
(opening)

sigmoid and rectum removed

Bowel Opening Through Abdominal Wall

A **Colostomy** is an operation in which the large bowel (colon) is brought onto the abdominal wall and opened. The stool (bowel movement) is then permitted to come out through this opening rather than through the rectum. After the operation, normal eating can begin 10 days to two weeks later. Return to a normal work schedule and sexual relations usually begin after 3 months.

Polyps of the Colon and Rectum

**Polyps
Can Lead
To
Cancer**

Intestinal polyps of the colon and rectum are common benign abnormal formation of tissue. They serve no useful function, but grow at the expense of the healthy organism. Thus it is called a _neoplasm_ (a thing formed).

Polyp is a word which in the Greek means _many feet_. A polyp is a tumor with a foot or stem. A polyp bleeds easily and sometimes causes painless rectal bleeding.

Polyps may be single or multiple and occur most frequently in the lower part of the descending colon (the _sigmoid_) and the rectum. The incidence of polyps increases with age.

Symptoms which may indicate polyp condition include:

1. Change in bowel habits
2. Blood in the stools
3. Increasing constipation
4. Anemia

In advanced stages there is generally a rapid loss of weight and anemia. These wartlike growths can develop into cancer.

It is generally believed that if there are multiple polyps in the colon and they are over 1/2" in diameter, the colon is potentially cancerous.

Methods Of Removal

Removal of polyps is done with a colonoscopic instrument that works inside the bowel. No external surgical incision is nessary. Small polyps can also be removed by electrical destruction of the tissue. This is called fulguration.

Colon Removed

Familial polyposis is a condition of multiple polyps that can occur in family bloodlines. Medically, it is believed that such persons with these multiple polyps are at a high risk of developing cancer.

Medically, the usual treatment in these cases is to remove the entire colon. The bowel opening is then made through the abdominal wall. This is called an *ileostomy* (groin opening). In this operation, a surgical passage is made through the abdominal wall into the ileum. The ileum is the lower three-fifths of the small intestines.

7

CANCER OF THE COLON and RECTUM

Colon Cancer Takes Toll

In medical terminology this is called <u>Colorectal Cancer</u> (colon and rectum). More than 100,000 new cases of cancer of the colon and rectum are diagnosed each year in the United States.

Cancer of the Colon and Rectum causes more deaths than any other form of cancer! This should lend some substance to the statement made at the beginning of this book that:

<u>*Death begins in the colon!*</u>

Males are affected more frequently than females in a ratio of 3 to 2. It occurs mostly in persons over 50 years of age. In Colon cancer the incidence is about as follows:

16% in the cecum and ascending Colon
5% in the transverse Colon
9% in the descending Colon
50% in the rectum

Many Within Reach

Almost two-thirds of the lesions of the colon and rectum lie within the reach of the examining finger or sigmoidoscope and therefore can be biopsied on the first visit to a doctor.

Symptoms

Symptoms of cancer of the Colon include:

1. Change in bowel habits
2. Blood in the stool (bowel movement)
3. Cramp-like abdominal pain
4. Short bouts of diarrhea
5. Jet-black stools

Surgery is the medical approach to cancer

COLON CANCER

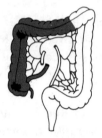

11%
Cecum and ascending colon

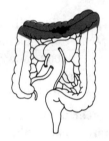

4.5%
Transverse colon

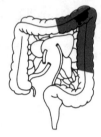

3%
Splenic flexure

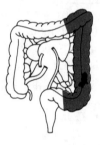

5%
**Descending colon
and upper sigmoid**

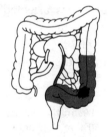

20.5%
**Low sigmoid
and upper rectum**

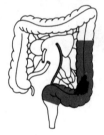

52.9%
Rectal

Chart indicates approximate incidence of cancer of the colon in each of six segments. Also shown are areas where cancer usually occurs.

of the Colon. Half the Colon may be removed without a great change in bowel movements. Or, in many cases, it is necessary to remove the entire Colon. Then the bowel movement is diverted to flow, not through the rectum, but to the outside through a surgical opening in the skin of the abdomen (colostomy). Bowel movements then collect in a bag.

In some cases x-ray or chemotherapy follows surgery. The overall 5-year survival rate after surgery resection is about 50% to 70% depending on whether the cancerous lesion was confined to a local area.

Two Treatment Methods For Cancer Of The Rectum

The Rectum is the lower part of the large intestine or Colon. It is about 5 inches long and connects to the anus, which is the outlet of the Rectum. About 25,000 cases of cancer of the Rectum appear each year in the United States.

Two new methods of treatment in rectal cancer include: supervoltage radiation and local excision. Supervoltage radiation is considered equal or superior to surgery. Also by this method of treatment the rectum is preserved and no colostomy is needed.

It must be noted that cancer of the Colon is rare in African people who have high-fiber diets and thus have bulky stools. Low-residue diets are common in the United States. This leads to stagnation of bowel contents and buildup of cancer-causing bacteria.

8

FISSURES, FISTULAS and FLATULENCE

Fissure

**Breaks
In
Skin**

Fissures in the anus region are small breaks in the skin in the area of the rectal outlet. These fissures are usually caused by constipation and with passage of large, hard stools that split the skin.

Medically, stool softeners are recommended and an anesthetic ointment which is applied to the anal canal both before and after a bowel movement. Healing generally takes place in two to three weeks.

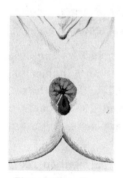

Fissure directly below anus

Chronic fissure problems are usually characterized by:

1. Acute pain during a bowel movement
2. Spotting of bright red blood while having a bowel movement
3. Tendency towards constipation

Fistula

**Abnormal
Tunnel**

An anal fistula is an abnormal connection between loops of bowel, or between the bowel and skin around the rectum, through which fecal matter discharges.[1]

---

[1]Donald G. Cooley, Better Homes and Gardens After-40 Medical Guide (Des Moines, Iowa: Meredith Corporation), 1980, p. 229.

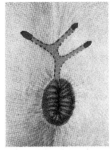

Fistula originating at anus. Note three external openings. Canal routes indicated by dash lines.

**More
Common
In
Youth**

Surgical probe indicates fistula canal leading to anal wall.

Note fistula canal size in relating to finger.

**Hardened
Stools
Serious**

Fistulas represent the end result of an infection that originated in the rectal area and which has tunneled its way out to the skin surface. The initial symptom is usually a painful boil or abscess alongside the rectum which opened and discharged pus. If not treated, this abnormal channel tends to spread and tunnel about the rectum and the sphincter muscle could be damaged.

The _sphincter muscle_ is a circular muscle that closes the anus.

Medically, a fistula in the anus will never heal spontaneously. It must be excised and the source of infection must also be removed, otherwise the fistula will recur. Sometimes when there are multiple fistulas, a temporary colostomy is performed. A patient could spend as much as six weeks in the hospital for proper healing to take effect . . . although the usual stay is 4-7 days.

Complete recovery can take one to four months. Fistulas are more common in young adults than in children and older people.

Fecal Impaction

Because of poor diets eating primarily low-residue foods, elderly people sometimes develop hardened or putty-like stools in the rectum and colon. This is particularly true of bedridden senile patients.

**Loss
Of
Control**

In many cases there is not only <u>fecal impaction</u> but also loss of control of bowel movements. This same condition can happen to young children who don't want to take the time to go to the bathroom.

The primary function of the colon is to extract water from food residue. Consequently, the longer the stool (fecal material) stays in the colon, the drier and harder it becomes.

The hard, cement-like mass forms a ball that sits firmly just above the anus. It can become very large in size. The individual may have a liquid type bowel movement because the liquid stool finds its way over and around this solid mass. Yet the individual is severely constipated and in danger of a complete bowel obstruction.

**Manual
Manipulation
Needed**

The impacted stool must be broken up manually. The nurse or doctor puts on a rubber glove and inserts one or two fingers in the rectum to break up this solid mass and allow the pieces to be expelled.

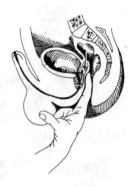

Cleansing enemas are given. Initially an olive oil retention enema followed by a warm water enema are the usual medical approach. In some cases, daily oil retention enemas are reccommended.

Flatulence (Tympanites)

Sources Of Gas

Flatulence is excessive gas in the stomach and intestines. Intestinal gas comes from air swallowed unavoidably during eating and drinking. Intestinal gases also come from gases in foods and from the action of colonic bacteria.

Flatulence may also be due to functional and organic disease of the digestive system. Some of this gas is normally absorbed from the intestines . . . while the remainder is expelled as <u>flatus.</u>

The gas that enters your stomach is just about pure air. As the air progresses down into your colon it can mix with other volatile products of fermentation including hydrogen sulfide. It is the hydrogen sulfide that gives the gas a rotten egg smell making the passing of gas offensive.

Eating too rapidly or eating under stress, drinking large quantities of liquids with a meal or chewing gum . . . all encourage intake of air and eventually gas in the intestines.

Milk products, carbonated beverages and foods in the bean family may lead to excessive gas in the colon.

Can Be Health Problem

Flatulence may be a symptom of a health problem. Excessive flatulence may be a symptom of gallstones or asthma and in some cases, heart disease.

There are some spices that increase the flow of digestive juices and check the growth of yeast and other gas-producing fermentation and may prove beneficial. They include: anise, cinnamon, cloves, lemon peel, mustard, nutmeg, onion, peppermint and thyme. Charcoal tablets (such as found in health food stores) are also beneficial to absorbing these gasses.

9

THOSE HUMILIATING HEMORRHOIDS!

**Dilated
Blood
Vessels**

Hemorrhoids are dilated blood vessels *(varicose veins)* both inside and around the rectal opening. In today's fast-paced life and low-residue eating habits, the individual finds himself straining when having a bowel movement.

In the process of moving one's bowels, the rectum is dilated (widened) so the stool can pass through. This repeated stretching of this canal takes its toll on the thin walls of blood vessels in that area. These veins can become weakened and permanently stretched and begin to bulge. Such a cluster of bulging veins are called Hemorrhoids.

**The
Curse**

Israel, during the day of Moses, was told that blessings would be theirs if they obeyed God. But if she disobeyed, many curses would befall her. These are outlined in the Old Testament in Deuteronomy 28:15-68. Among these curses are boils and hemorrhoids (Deuteronomy 28:27).

The Painful Plague

When the Philistines captured the Ark of God and took it to the temple of their idol Dagon in the city of Ashdod .. trouble began. The Lord caused a plague of hemorrhoids to come upon them.

> ... He (God) smote the men of the city
> both small and great,
> and they had hemorrhoids
> in their secret parts ...
> and the cry of the city
> went up to heaven.
>
> (1 Samuel 5:9, 12)

Anyone who has had hemorrhoids knows that they can be very painful and embarrassing.

**Many
Suffer**

Perhaps one-third of all adults at one time or another suffers from hemorrhoids.

Hemorrhoids are classified as either internal or external. An individual can have both at the same time.

Internal

Internal hemorrhoids are situated well up in the canal above the *sphincter* muscle.

Internal hemorrhoids are multiple, soft, purple, irregular in shape and covered by a thin layer of mucous membrane. The principal symptom of internal hemorrhoids is bleeding. If one has a sense of an incomplete bowel movement the presence of large internal hemorrhoids may be indicated. If straining at the stool becomes a habit, there is a danger that these hemorrhoids may prolapse . . . that is, a falling or dropping down through the anus.

External

External hemorrhoids appear as small rounded purplish masses covered with skin. They become more prominent when one strains while having a bowel movement. Generally they are soft and not painful. If they enlarge developing a blood clot, the condition can become very painful for about 5 days until the clot is absorbed.

**Common
Cause Of
Bleeding**

Hemorrhoids are the most common cause of rectal bleeding. The blood is bright red whereas bleeding from the stomach or small intestines is usually a black tarry color.

There are two important symptoms of internal hemorrhoids: <u>bleeding</u> and <u>prolapse (falling or slipping out of place in rectum)</u>.

<u>Bleeding:</u> The first sign may be a slight streak of blood on the toilet paper, especially if the individual is constipated. Later, there may be a steady drip of blood for a few minutes after the bowel movement is passed. At a more critical stage, the bleeding may occur at any time in the day apart from having a movement.

<u>Prolapse:</u> This is a later development and it occurs at the time of having a bowel movement, initially. The hemorrhoids or piles appear at the anal opening and slip back after a movement. Later, the piles tend to remain prolapsed outside the anal canal and have to be pushed back by hand into the anal canal. Eventually, they continue to remain outside the canal.

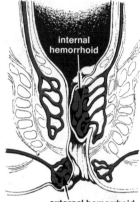

internal hemorrhoid

external hemorrhoid

<u>TREATMENT</u>

There are three forms of treatment medically:

1. **<u>Palliative</u>**
   Change in diet is recommended. Use of ointments and suppositories may be suggested.

2. **<u>Injection</u>**
   In some cases, a weak carbolic solution is injected not in the pile itself but above it. This is usually an outpatient procedure and requires no special preparation of the patient or the bowel. There is little if any pain.

3. **<u>Surgery</u>**
   A. <u>Rubber-band treatment</u>
      The doctor grasps the hemorrhoid with a special device that makes it possible for him to slip a rubber band very tightly over the base of the hemorrhoid. The tight band shuts off the blood supply of the hemorrhoid which sloughs off in a few days. There may be some pain.

B. Hemorrhoidectomy

This is surgical removal of the hemorrhoids. It is done under general anesthesia or a low spinal although a local anesthesia is sometimes given. The operation takes about 20 minutes. Retention of urine after a hemorrhoidectomy is sometimes experienced. Normal bowel functions may not return for several weeks and there may be some evidence of bleeding in the stools for a few days or a few weeks. There is some post-operative pain particularly with the first few bowel movements immediately after surgery.

C. Cyrosurgery

Freezing (cyrosurgery) is the application of extreme cold to the hemorrhoids. It has been used since 1969.

A foot-long probe is attached to a reservoir of liquid nitrogen. This keeps the tip of the probe at a temperature some hundreds of degrees below zero.

When the tip of the probe touches the hemorrhoid, it freezes the tissue instantly and painlessly. Nerve fibers of the hemorrhoids are destroyed, eliminating any sensation. Surrounding tissue is not affected.

The frozen-dead hemorrhoidal tissue puffs up with fluid that drains for about 3 or 4 days. It is necessary to wear an absorbent pad. This procedure takes only about 15 minutes and can be done in a doctor's office.

Normal activities can be returned to within one day.

External hemorrhoids are usually treated with Sitz baths and stool softeners. If surgery is required, a local anesthetic is usually applied to the hemorrhoid area. A small incision is made so the blood clot can be squeezed out. A pad is placed over the area until healing is complete.

For mild cases of hemorrhoids, medical doctors usually suggest suppositories that contain hydrocortisone. Often a <u>sitz</u> bath is recommended. This is where the water in the bathtub covers the hips. The water should be hot.

Hemorrhoids And Piles

If the hemorrhoid veins dilate but remain up in the anus . . . they are called <u>*hemorrhoids*</u>. If they protrude like a pillar or tower, they are called <u>*piles*</u>.

Chronic hemorrhoids enlarge and repeatedly cause severe problems. Doctors recommend surgery to correct this problem. In the past surgical treatment included amputation of the hemorrhoid, chemically injecting it, freezing it and cauterizing it.

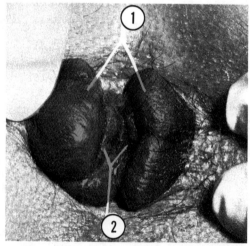

Photograph shows internal hemorrhoids that have prolapsed . . . slipped out of the rectum. (1) The inflamed tissue of (2) the hemorrhoids.

Currently <u>rubber band</u> therapy is widely used. A rubber band is wound tightly around the hemorrhoid causing circulation of blood in that area to cease. The protruding hemorrhoid dies and the problem is resolved.

More common, however, is surgery where the surgeon actually cuts out the veins. This is done either with a local or general anesthesia. It is called a *hemorrhoidectomy*. The operation takes about 20 minutes. The hospital stay is 4-7 days. For the first week or two after the operation there is pain in that area. It may take several weeks before normal bowel function returns.

May Recur After Surgery

Lubricating laxatives are given as well as frequent sitz baths. A bowel movement occurs on the 3rd or 4th day. Because of the post-operative pain involved, many

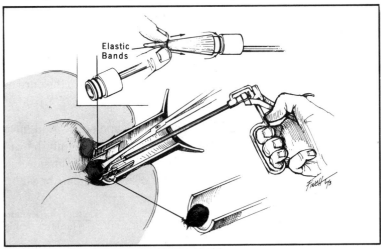

Elastic Bands

Rubber band is placed around hemorrhoid.

become disenchanted with surgery. This is particularly so because such surgery is no guarantee that hemorrhoids will not recur.

Hemorrhoids are rare in countries such as Africa where the diet includes a high fiber content (high residue). Wherever American or European influences are noted, such as in southern Nigeria, hemorrhoids are nearly as common as in Western countries.

Diet Change Needed

Except where hemorrhoids have caused structural changes in the rectal area . . . the simple addition of fiber to the diet can relieve symptoms of hemorrhoidal problems.

Hemorrhoids do not cause cancer, and they do not need to be removed or treated *"before they turn into something."* But hemorrhoids and cancer may occur in the same general area; they may imitate each other by bleeding, and therein lies the danger . . . . Competent examination is the only means of accurate diagnosis, and early diagnosis is absolutely vital to the cure of rectal and colon cancer.[1]

One still wonders if Napoleon would have won the battle of Waterloo if only he did not have hemorrhoids. You might say that is where the seat of his problems began!

[1]James Ferguson, M.D., The Inside Story of Hemorrhoids (Indianapolis, Indiana: The Saturday Evening Post), 1981, May/June, p. 81.

10

THAT DISTRESSING DIARRHEA!

**Causes
Of
Diarrhea**

Diarrhea means, in the Greek, *to flow through*. And this definition is quite descriptive of the problem. For in diarrhea, there is a frequent passage of watery bowel movements.

About 90% of diarrhea problems are temporary and soon pass away. There are at least 11 known causes of diarrhea.

1. Emotional
   This is a common cause. Preparing to get married, getting ready to take a trip, or an emotional upset can trigger a set of digestive responses that brings about diarrhea. This is a primary cause.

2. Intestinal infections
   Food poisoning, dysentery, staphylococcal infections, fecal impaction, antibiotic therapy, inflamatory bowel disease or cancer.

3. Malabsorption
   Celiac sprue involving impaired absorption of fats, glucose and vitamins.

4. Pancreatic disease
   Pancreatic insufficiency or tumors on pancreas.

**Causes
Of
Diarrhea**

5. Cholestatic syndromes
   Stoppage of bile excretion.

6. Neurologic disease
   Diabetic neuropathy, syphillis affecting the spinal cord.

7. Metabolic disease
   Excessive secretion of thyroid glands

8. Malnutrition
   Diarrheic diseases of infancy are most common from 6-18 months of age. Child has restless sleep, general malaise, abdominal pain and headache. Urine is scanty and the anus is chafed and sore from urinal acidity. Heart becomes weak. When a child develops persistent diarrhea your physician should be contacted immediately. This form of diarrhea is called *Marasmus.*

   Kwashiorkor is another form of malnutrition found in countries like Africa where famine abounds. It is a lack of protein in the diet.

9. Food allergy

10. Dietary factors
    Changing from a heavy meat diet to a nutritional diet of faw foods and juices can initially bring on a period of diarrhea. Many of us have experienced *"travelers diarrhea."*

11. Drug-induced
    Many drugs can cause diarrhea as a side effect. *Meperidine hydrochloride* (Demerol), *phenylbutazone* (Butazolidin), *indomethacin* (Indocin) and many other drugs can cause diarrhea.

**Food
Not
Absorbed**

In diarrhea, very little of the food eaten is absorbed. Actually, you are losing more water than you have taken into your system. With cramps and pain, you swallow more air and eventually produce gas. With the loss of so much water, there is a danger of dehydration. In severe cases, there is also vomiting. This is called gastroenteritis.

**Nutrients
Lost**

In diarrhea you lose vitally important minerals and salts, especially potassium salts.

> Lack of potassium salts in the body has the effect of making you tired and listless and weak. Potassium is necessary for the smooth functioning of all muscles. In extreme potassium lack, the heart muscle itself can stop functioning.[1]

Potassium, along with certain other salts including sodium are called electrolytes. The reason . . . when they are dissolved in water, they release atoms (called ions) which are capable of carrying electric charges. Electrolytes are either acids or alkalies. In diarrhea, the intestinal contents that are lost are alkaline. This alkali loss results in a condition called acidosis.

**Severe
Diarrhea**

In severe diarrhea then, problems can develop in three different areas:

1. Loss of water
2. Loss of potassium
3. Upset of acid-alkali balance

In acute bacillary dysentery or cholera, such a condition can cause the individual

[1]Marian T. Troy, M.B., Better Bowel Health (New York: Pyramid Books) 1974, p. 47.

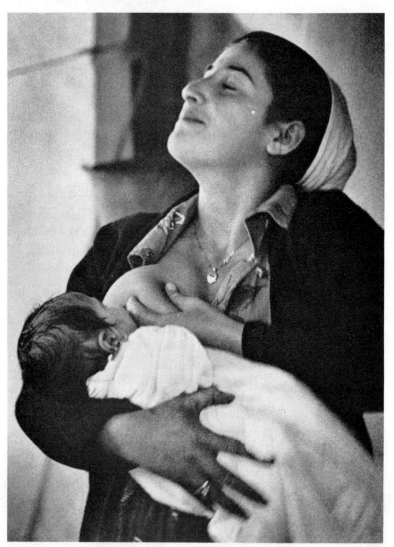

Bottle-fed babies are more prone to develop diarrhea. Mothers who breast feed their babies pass on to the infant important bacterial antibodies. Because of this breast-fed babies enjoy better health and a better start in life. This photograph taken near Jerusalem in 1960 reveals some of the emotional values that both mother and child enjoy through breast feeding.

to die in two days. This is why so many died during wars where soldiers were imprisoned in crowded camps and malnourished. This also occurs in areas of the world where there is extreme famine.

Infant Diarrhea

Can Become Critical

In an infant, diarrhea can become very serious. The loss of just 2 ounces of water in a 7 pound baby is equivalent to the loss of about 2 quarts of water in an adult weighing about 170 pounds.

Breast Feeding— The Ideal

Fortunately, mothers who breast feed their babies pass on to the infant important bacterial antibodies. Bottle-fed babies are more prone to develop diarrhea. Only human milk is tailored to meet the complete needs of the infant. Then, too, there are no middle men and distribution chains. Breast milk goes directly from the producer to the consumer and is readily available on demand.

Babies occasionally develop diarrhea because of dietary changes or adjustments. If the number of stools doubles in a day and the child looks and acts ill, it is important you contact your doctor.

Green Stools

If your infant passes green stools it is an indication that the intestinal contents are rushing through the infant and the bile has not been working effectively. The infant usually is crying and nothing seems to pacify him ... not even holding him in your arms. He becomes pale, rings around his eyes, his hands and fingers turn from pink to white. He may have developed a dangerous form of diarrhea called _enteritis_. If this is true, he should be taken to the hospital immediately.

A normal infant, taking nothing but milk doubles his birthweight in 4-1/2 months. He triples his birthweight in 12 months. There is a tendency of mothers (and grandmothers) to begin to overfeed the child. When this occurs, diarrhea can follow.

Too Much Fat

If the child is getting too much fat in his formula, the stools will smell and will float on top of the water in the toilet. Your physician will most likely recommend that you cut down on the amount of fat in the formula.

Too Much Starch

If the infant's formula contains too much starch and sugar, the stools will smell sour and will look frothy. It is an indication of too much carbohydrate in the formula for that infant.

Underfeeding can also result in diarrhea in infants. The commonest reason is that the holes in the nipple of the formula bottle are too small. The child strains, swallows a lot of air and goes hungry. Naturally the infant will become collicky.

Lomotil or similar drug is sometimes prescribed by doctors for infants. Never exceed recommended dosage. Lomotil is not for use in antibiotic-induced diarrhea. It can cause loss of appetite, nausea, vomiting and drowsiness.

A nipple with too big a hole can cause the same problem. Getting too much formula, he chokes and limits his intake.

If at all possible, your child deserves the best . . . and that is breast feeding. Commercially-prepared formulas are homogenized, sterilized mixtures of soybeans, conditioners, freshness preservers, synthetic vitamins and minerals and are a poor substitute for the real thing.

The Vote

On May 20, 1981, the World Health Organization (WHO) voted to encourage women around the world, especially in poor nations, to breast-feed their children rather than give them manufactured substitutes. Only the United States voted against the measure. Proponents of breast-feeding reported that high-powered sales tactics discourage breast-feeding. In Third World countries, formula is mixed with polluted water under poor health conditions contributing to millions of infant deaths. Breast-feeding is regarded as the healthiest infant nourishment.

Travelers Diarrhea

Travelers Diarrhea generally starts out with constipation. This comes on because of inactivity, tiredness and the tendency to overeat.

Then comes the diarrhea and thirdly, a stomach upset may occur.

Generally doctors recommend commercial preparations such as Kaopectate, Kalpec, Atasorb or activated charcoal to relieve these symptoms. If these do not work, a prescription drug such as Donnagel or Kaomycin may be recommended.

Chronic Diarrhea
(Chronic Enteritis)

Emotional Upset

Many cases of chronic diarrhea are due to anxiety-producing mechanisms. Periods of depression and anxiety as to one's financial security and extreme fears of being alone can bring on this chronic condition. Such conditions will not be resolved by simply taking diarrhea medications.

Chronic diarrhea may also come about because of the abuse of laxatives.

Functional disorders (irritable bowel) or disease of the colon and small intestine (cancer, ulcerative colitis, Crohn's disease) can also cause chronic diarrhea.

The physician must determine the root cause of chronic diarrhea in a patient before arriving at a mode of treatment.

THE CURSE OF CONSTIPATION

CONSTIPATION *(To Press Together)*

Two Approaches

Constipation is at the opposite end of the spectrum of intestinal problems. In diarrhea, there is an abnormal flow, in constipation one feels as though the Grand Coulee dam has suddenly closed its gates.

There is a wide variance between most medical doctors and nutritionists as to what constitutes constipation.

Those in orthodox medicine feel that in some, a bowel movement every 2 or 3 days can be normal. Nutritionists believe that a normal bowel movement should occur about twice a day and preferably after each meal.

Two Categories

Constipation is generally categorized into two types:

1. Atonic *(not stretching)*
   In this form, the muscles of the colon have lost their tone and are unable to contract and relax properly to propel bowel wastes. This usually affects the very beginning of the colon in the area of the cecum.

2. Spastic *(convulsive)*
   In this form, the colon may react with spasm-like effect. The food residue, therefore, does not pass along as it should.

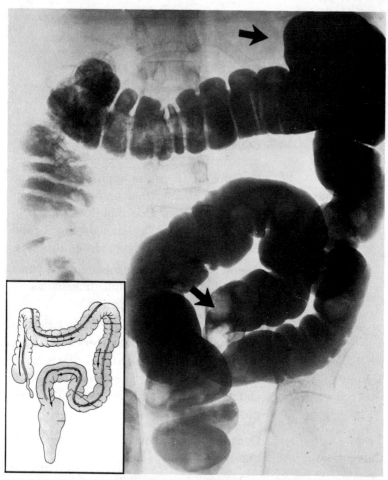

This is an x-ray of a colon of a male, age 57. Arrows point to kinks or twistings in colon. Such a condition leads to bowel problems. Inset shows normal colon shape.

The colon can become atonic (losing its muscle tone) in one area and spastic in another area.

One is considered constipated if bowel movements are delayed for a number of days or if the stools are unusually hard, dry and difficult to pass. Constipation can mean infrequent stools, straining while attempting to move your bowels, or a feeling of incomplete bowel action.

Symptoms Of Constipation

Symptoms of constipation include a general tiredness, bad breath, fatigue and lack of zest. Constipation is sometimes referred to as following:

1. Simple constipation

   In simple constipation, it is generally related to diet. There is a tendency in developed countries like the United States and countries in Europe to eat a large proportion of highly refined foods. These provide very little residue for the intestines to work on. They also encourage illness in later years.

   Environmental factors can encourage simple constipation. Flying long distances, pilots and flight attendants are prone to this ailment. Drugs taken for high blood pressure can encourage simple constipation.

   To correct constipation . . . raw foods are encouraged with high fiber content . . . salads and fruits. Refined breads, sweets, fried foods and the traditional hamburger on a roll simply encourage constipation and should be avoided.

2. Severe constipation (*obstipation*)
   This is a less frequent cause of constipation and usually is attributed to
   psychiatric disorders or as a result of
   medical or surgical treatment. It can
   come about because of fecal impaction. It can also be a sign of cancer of
   the right side of the bowel.

**Change
In
Diet**

Medically, constipation is treated with a
suggested change in diet that allows more
high fiber content and raw fruits and
vegetables. Six to eight glasses of fluid are
recommended daily as well as regular exercise.

Enemas are occasionally used as a temporary expedient. Generally a saline enema
is indicated. Sometimes an oil retention
enema is used. This is placed in the rectum
in the evening and retained overnight and
evacuated the following morning.

A coffee enema flushes out all the loosened fecal matter in the colon and stimulates the liver to throw off much of its toxic wastes, according to some nutritionists.

Dr. William D. Kelley, who has been involved in cancer research for over 20 years, states, that to detoxify his patients:

We use three things:
coffee enemas,
a purge and fast,
and a liver and gall bladder flush.

The most important is the daily coffee enema.
This stimulates the liver
    to throw off much of its toxic wastes.

Coffee enemas should be started
even before you start a nutritional program
because the body cannot begin to rebuild itself
while the old poisonous waste products
still clog the tissues.

Moreover,
someone on the nutritional program
who is not taking regular coffee enemas
would most certainly get sick.
The body would be trying to throw off
enormous amounts of accumulated poisons,
so these poisons would come out of storage.
Without the enemas,
    most of them would remain in the blood
    and make the person quite sick.[1]

In making a coffee enema preparation, do not use aluminum or teflon utensils. Dr. Kelley suggests they never be used at any other time, either.

Coffee enemas are best taken in the morning or early afternoon. Taking a coffee enema at night may keep you awake. It is best to try to have a bowel movement just prior to your taking a coffee enema.

The coffee you use for a coffee enema should be the **non**-instant variety. It should also be **non**-decaffeinated coffee.

---

[1]Sam Biser, The Healthview Newsletter, Vol 1, Number 5, p. 7. (A complete transcript of this report can be secured by sending $2 direct to: The Healthview Newsletter, Box 6670, Charlottesville, Virginia 22906)

Dr. Kelley suggests that you use <u>2 to 4 tablespoons</u> of coffee grounds <u>per quart</u> of distilled water. V.E. Irons recommends that you put about 12-15 tablespoons of coffee into a <u>gallon</u> of water . . . boiling it for 10 minutes. He suggests a gallon because *". . . you might have to start over again if you can't retain the enema."*

Both suggest taking a warm water enema first to remove the gas and large particles of residue. Then follow this with a coffee enema.

The coffee enema preparation can be made the night before and allowed to sit overnight so it is body temperature.

Put two quarts of coffee into an enema bag. The coffee should be luke warm. If it is too hot or too cold, your colon muscles will react and expel it.

Dr. Kelley suggests the use of a colon tube which is 30 inches and reaches most of the colon. V.E. Irons states that you don't need a long colon tube.

In the colon tube therapy, Dr. Kelley states that the colon tube should be lubricated with K-Y jelly or some other suitable lubricant.

> *Insert the tube 18 to 20 inches into the rectum,*
> *rotating it to avoid kinking the tube.*
> *If you hit an impassable area of the colon,*
> *of course, stop there,*
> *as some people can't insert the tube the full length.*
> *If the coffee does not run in,*
> *the tube has kinked.*
> *Pull it out until the flow of solution is felt.*[1]

An enema should <u>never</u> be taken while standing or sitting on the toilet. Dr. Kelley suggests you lie five minutes on your left side, five minutes on your back, then five minutes on the right side . . . then empty yourself.

V.E. Irons suggests, after the enema tube is removed, that you kneel in the standard enema position . . . on all fours with your head and shoulders low and your buttocks up. He suggests you rest your body weight on your knees and on one hand. In this position, the colon drops downward. This makes it easier for you to gently massage the area. Irons suggests you start massaging a 2-inch area of the colon on the lower left side of your abdomen and then gradually move

---

[1]Ibid., p. 7.

into other areas of the colon. The purpose of the gentle massage is to loosen any fecal deposits which are densely impacted.

The section of bowel from the middle of the abdomen to your lower right-hand side usually contains the greatest amount of encrusted fecal matter, V.E. Irons states.[1]

A coffee enema can take as long as an hour. V.E. Irons recommends that coffee enemas be taken each day you are on a cleansing program.

It is wise to remind the reader that no enemas should be taken without the approval of your doctor.

---

[1] Ibid., Number 10 Healthview Newsletter, p. 4. V.E. Irons interview.

ALL ABOUT ENEMAS

In Louis XIV's day, enemas were a fad. Many women in royalty had enemas three times a day. In fact the King himself had 107 assorted doctors and in one year they administered to the King 215 enemas.[1]

Enema means *"to throw in"* and medically they are usually given by injecting a liquid solution into the rectum and colon to empty the lower intestine or to introduce food or medicine for therapeutic purposes. A regular enema reaches only about 5 or 6 inches of the colon.

A **high enema** is designed to reach the colon. Thus a long rubber 30 inch tube is extended into the rectum to carry the cleansing liquid as far up as possible in the colon.

Basically there are **four** types of enemas:

1. Barium enema
   Administration of barium sulfate in solution as a diagnostic aid in X-ray examination of the colon.

2. Carminative enema
   One given to relieve distention caused by flatus (gas) and to stimulate peristalsis.

3. Cleansing enema
   One to empty the lower intestine or the colon. Often a coffee enema is used for this process.

4. Retention enema
   A coffee enema is considered a retention enema in that the liquid is retained in the colon for a period of time. Retention enemas are used also to soothe or lubricate the rectal mucous membrane, to apply absorbable medication or to soften feces.

How To Take An Enema

To take an enema, you must have an enema can or bag with a rubber hose and a nozzle; it can be obtained at any drug store.

Fill the enema bag with lukewarm water, about 99 degrees F. Add a few drops of fresh lemon juice, or a cup of camomile tea (can be bought at health food stores); however, the enema can be taken with plain water. For a do-it-yourself enema, 1 pint to 1 quart water is sufficient.

---

[1]Marian T. Troy, Better Bowel Health (New York: Pyramid Books), 1974, p.180.

The best position for taking an enema is on your knees, head down to the floor, with enema bag hanging 12 to 18 inches above the anus, to get sufficient pressure in the flow of water. The flow can be regulated by squeezing the tube with the fingers; some enema bags have a special clamp to regulate the flow.

Before inserting the nozzle into the anus, make sure there is no air left in the tube; let water run for a moment. Use some vaseline, oil or other lubricant on the nozzle to make insertion easier. If you feel discomfort or pain when water is running in, stop the flow for a while and take a few deep breaths; then continue again until the bag is empty.

If you can retain the water for a while and do not feel forced to empty the bowels at once, you may lie on a bed or soft rugs for a few minutes and let the water do its dissolving and washing work before letting it out. First lie on the back for a minute, then on the right side, then on the stomach, and then on the left side. While you are doing this, gently massage your stomach with your hands. Then go to the toilet and let the water out. Stay long enough to make sure the bowels are empty.

For a small child or a baby, a small all-rubber-bulb ear syringe can be used. The tip of the bulb should be lubricated with petroleum jelly. Do not use soapy water. For an infant, use up to 4 ounces of luke warm plain water. For a 1-year-old, up to 8 ounces of water. For a 5-year-old, up to one pint. Gently insert the lubricated syringe tip into the anus. The slower you inject the solution, the more effective the enema will be. Hold the child's buttocks together to prevent a premature evacuation.

Enemas are preferred over laxatives because there are fewer dangers of sensitivity, allergies and interference with normal bowel functions. Diabetics do not have to worry about sugar nor the heart patient about sodium (as they would if they took a laxative) because an enema only reaches the colon.

An enema has the advantage in that it acts quickly . . . within a few minutes. Enemas, too, have a soothing effect and are safer to use in cases of constipation than laxatives.

12

HELP WITH HIATAL HERNIA

Low Residue Diet May Be To Blame

The hiatus is a normal hole in the diaphragm. The diaphragm is the muscle of breathing and is through which the esophagus joins the stomach. If its supporting ligaments become so stretched and weakened that abdominal pressures force a protrusion of the junction into the chest . . . this becomes a *hiatal hernia*.

It is interesting to note that in England about one-third of the population suffers from hiatal hernia. Many believe low-residue diets are to blame. And in areas where constipation is a part of life there is a higher incidence of hiatal hernia.

When constipated, tremendous pressures are exerted on the colon. This causes the esophagus to distend. This repeated action over the years weakens the muscle fibers and a bulge appears in the esophagus or the diaphragm above the stomach.

Obesity, chronic coughing or giving birth to several babies over the years can also cause this problem.

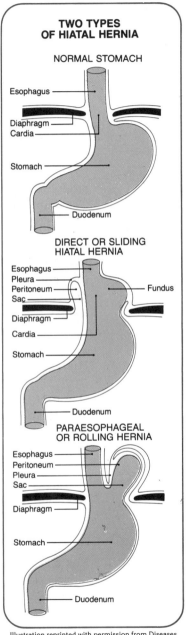

TWO TYPES OF HIATAL HERNIA

NORMAL STOMACH

Esophagus

Diaphragm
Cardia

Stomach

Duodenum

DIRECT OR SLIDING HIATAL HERNIA

Esophagus
Pleura
Peritoneum
Sac

Diaphragm

Cardia

Stomach

Fundus

Duodenum

PARAESOPHAGEAL OR ROLLING HERNIA

Esophagus
Peritoneum
Pleura
Sac

Diaphragm

Stomach

Duodenum

Symptoms

Symptoms may include:

1. A burning pain under the breastbone
2. Regurgitation of stomach contents into lower esophagus
3. Feeling of fullness which begins soon after eating
4. Belching or hiccupping

Symptoms are most noticeable while straining or stooping and when reclining. The pain sometimes mimics a heart attack. It is a problem which occurs in up to 50% of the U.S. population after age 40.

Medically, it is treated with antacids and the eating of six small meals a day rather than 3 large meals. If there is severe and repeated problems surgery is performed.

Hiatal Hernia Exercise

A
Possible Aid

One exercise many chiropractors suggest to alleviate hiatal hernia is the following:

Lie flat on the floor
with arms straight over the head.

Slowly sit up,
keeping the arms straight up
over the head
and keeping the back straight.

The feet should be wedged
under a chair
to hold them on the floor.

Then slowly
lie back down . . .
keeping the arms and back straight.

Do this two or three times daily.

13

The Nutritional Approach To
BOWEL PROBLEMS

**The Name
Not Important**

As we said earlier in this book, nutritionists believe that

Death begins in the colon!

While the medical doctor approaches bowel problems seeking symptoms and then dispensing drugs or performing surgery on a particular problem ... the nutritionist looks at the whole body.

Knowledgeable nutritionists are not basically interested in the name of a disease but in the overall functions of the body.

As an example, treating hiatal hernia by antacids and small meals is the medical approach. But the basic root cause, for which the nutritionist centers his concern, is constipation.

**Unleashing
A Host
Of Diseases**

With poor functioning bowels a host of diseases can be unleashed in the human body from colitis, to gallstones, to ulcers, to cancer. Therefore, for the nutritionist the first step is to cleanse the colon and get it functioning normally again. When this is done, they believe, the body will begin its healing process throughout the system.

Look at the foods you eat. Ask yourself the question each time you sit down to eat:

> Is the food I am about to eat
> going to make my body
> a gentle, flowing stream
> or
> a stagnant pool?

If your daily intake of food is made up mostly of highly processed food, low residue food, cooked food . . . then your body is going to become a stagnant pool. You should keep your health insurance paid up for you are encouraging serious illness to occur.

It was Catharyn Elwood who wrote:

> We in America
> spend $200 million dollars a year
> to move our bowels!

> This is quite a price,
> especially when it can be done
> more efficiently—for free!

> The chief reason we need
> so much bowel "dynamite"
> is because we eat so many lifeless foods.[1]

Raw Foods Recommended

Catharyn Elwood recommends a diet high in raw foods. And if these cannot be tolerated, she suggests taking one tablespoonful of psyllium-seed preparation with whey or gelatine in a glass of water with each cooked meal.

Because the digestive tract has lost its muscular tone, the addition of B-complex vitamins to a diet is essential . . . especially Vitamin B_{12}. Not only do the B-complex vitamins help your bowels but they also calm your nerves.

[1]Catharyn Elwood, Feel Like A Million (New York: Pocket Books) 1976, p. 215.

THE SQUATTING
POSITION BEST

**After
Each Meal**

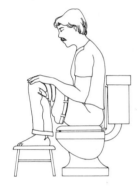

Catharyn Elwood suggests that you strive to maintain normal elimination by giving nature a chance to evacuate your bowels 20 to 30 minutes after each meal. She recommends a small stool (10″ to 14″ high) for your feet . . . so that the abdominal muscles can contract and relax normally and aid in the defecation.

The *"squatting"* position is the natural posture for easy evacuation. The reason a stool is suggested is because modern toilet devices do not permit the squatting position. This position strengthens the abdominal walls at their weakest point and eases the procedure of moving one's bowels.

PEELED GARLIC SUPPOSITORY

**Garlic
And
Hemorrhoids**

If there is tenderness, hemorrhoids or irritation at the anus, Catharyn Elwood recommends a peeled garlic bud, oiled and inserted as a suppository and allowed to stay overnight as a natural healing aid.

Catharyn Elwood received her Master's Degree in Food and Nutrition from the University of Maryland.

Mineral oil should never be used as a laxative. It robs the body of the fat-soluble vitamins (A, D, E, K) that are waiting to be assimilated in the intestinal tract.

YOGURT A FRIENDLY ALLY

**Yogurt
Beneficial**

Yogurt is very beneficial in reducing putrefaction.[1] Yogurt also improves bowel function. It helps eliminate flatulence (gas) as well as heartburn. Studies have shown that yogurt destroys the harmful intestinal bacteria yet sets up a powerful Vitamin-B-manufacturing factory in the intestinal tract. This B-complex keeps the intestinal tract clean, promotes better digestion and energy and contributes to beautiful hair and skin.

To be most beneficial, a cup of yogurt a day is essential. Don't buy yogurt which contains fruit or is flavored as this has sugar in it which destroys the effectiveness of friendly acidophilus bacteria.

RAW JUICES A LIFE SAVER

**Raw
Juices
Vital**

Chronic constipation can cause distention of the abdomen. In older people who suffer from chronic constipation there is a lack of vitality and their skin tends to become yellow and parchment-like. In younger people with this problem, acne and other skin eruptions occur.

Proper diet is essential to a healthy colon. About 60% of one's diet, according to nu-

[1]*Putrefaction* is decomposition of animal matter, especially protein associated with malodorous and poisonous products such as ptomaines, mercaptans and hydrogen sulfide, caused by certain kinds of bacteria and fungi.

tritionists, should be fruit and vegetables . . . of which 30% of this should be raw.

Some elderly people cannot tolerate raw foods initially and therefore a juicer is essential. Raw vegetables and raw fruits can be juiced and this liquid will provide important essential nutrients to the body.

Some adults cannot tolerate bran, which is an excellent laxative. The bran required in a diet, however, can be found in stone ground whole wheat bread. It is easy to assimilate and provides essential nutrients not found in commercial breads.

RUTIN RESOLVES CRISIS

Lecithin And Rutin

Hemorrhoids sometimes disappear when liquid lecithin is applied once or twice a day for several days. A lecithin capsule is opened and the contents applied to the hemorrhoidal area.

Bioflavonoids (Vitamin P) are water-soluble and often appear in fruits and vegetables as companions to Vitamin C. The components of the bioflavonoids include rutin. Bioflavonoids are essential for the proper absorption and use of Vitamin C. They increase the strength of the capillaries and help prevent hemorrhages and ruptures while building a protective barrier against infections. Some have found that applying rutin to painful and swelling hemorrhoids brought quick relief.

One individual who received heartwarming results with <u>rutin</u> took three 50 mg. tablets of rutin a day. In two day's time the swelling and pain were completely gone.

Other Aids

Vitamin B$_6$ after each meal has helped some people with hemorrhoid problems.

Others have made direct applications of 400 I.U. of <u>Vitamin E</u> three or four times a day and experienced cessation of itching and swelling.

Bioflavonoids are found in the white inner rind of oranges and other fruits. Some nutritionists recommend taking the entire spectrum of bioflavonoids, not simply rutin. When there are hemorrhoid problems, they suggest 12-15 tablets of <u>Bioflavonoid Complex</u> made up of:

Citrin	100 mg.
Hesperidin	50 mg.
Rutin	50 mg.

They also suggest 25 mg. of Vitamin B$_6$ after each meal and 1000 mg. to 2000 mg. of Vitamin C daily.

1
The WALKER Approach To Colon Health

Norman W. Walker, Ph.D. has for over 70 years practiced and preached sound nutrition. In 1910 he established the Norwalk Laboratory of Nutritional Chemistry and Scientific Research in New York. In the 1930's he was among the first to advocate the therapeutic value of fresh vegetable juicers.

Dr. Walker is a firm believer in **colonic therapy.** Walker believes much of the colon problems are concentrated in the <u>first</u> part of the colon which is 1½ feet long and called the *ascending colon.* From the *cecum* to the middle of the *tranverse colon* much of the accumulated fecal deposits line the colon wall . . . unreached by enemas or laxatives.

It is this first portion of the colon that is so critical for good health. The <u>ascending</u> colon extracts from the residue coming from the small intestine any remaining available nutritional material. These vital nutrients are collected by the blood vessels lining the walls of the colon and carried to the liver for processing.

If the feces in the colon have putrefied and caked themselves on the colon wall, polluted nutrients pass into the blood stream. According to Walker, this is called *toxemia.* He identifies <u>toxemia</u> as:

> . . . a condition in which
> the blood contains poisonous products
> which are produced by
> the growth of pathogenic, or disease-producing bacteria.
>
> Pimples, for example,
> are usually the first indication
> that toxemia
> has found its way into the body.[1]

Another important function of the <u>ascending</u> colon, according to Walker, is to gather from the glands in its walls the intestinal flora needed to lubricate the colon. Enemas and colonics do not wash out the intestinal flora but, rather, restore the colon to a healthy, functioning level.

Dr. Walker does **not** approve of <u>oxygen</u>-colonics. He believes oxygen should enter the body through the air we breathe and that the water alone when flushed through the colon provides natural oxygen.

---

[1]Norman W. Walker, Ph.D., <u>Colon Health</u> (Phoenix, Arizona: O'Sullivan, Woodside & Company) 1979, p. 13.

2
The WALKER Approach To Colon Health

Dr. Walker recommends initially at least 12 colonics in a series of treatments if one has had poor nutrition and bowel habits. Thereafter, he suggests colonics twice a year.

In his book, Colon Health, Dr. Walker indicates the relationship between the colon and a corresponding part of the human anatomy. He devotes a chapter each to the hypothalamus and pineal glands, the eye, the ear, the tonsils, the spine, the bronchials and lungs, the thyroid gland, the thymus gland, the heart, the stomach, the liver, the gall bladder, the pancreas, the spleen, the appendix, the testicles, the uterus and the mammary glands. He relates each of these to a specific place in the colon. His book contains a colon chart outlining on the colon where each gland or organ is affected by a malfunctioning colon and what biochemical cell salts are involved.

As an example Walker relates the eye to a position on the ascending colon just above the ileocecum area. He relates the mammary glands, genital glands, uterus and prostate in various positions in the rectum area of the colon.

Dr. Walker believes in cosmic energy vibrations. He states that "... vibrations cause or form energy, and energy is the result of vibrations."

Each part of the body and each gland, Walker suggests, has its own individual rate of cosmic energy vibrations. As ailments occur, vibrations are lowered or reduced.

> By checking the vibrations
> of each of the various parts of the afflicted body,
> the organ or gland actually involved
> could be readily pinpointed and corrected.[1]

Along with colonics, Dr. Walker is a firm believer in building up the connective tissues with ascorbic acid (Vitamin C). Because the body does not produce its own ascorbic acid, he is not a believer in prolonged fasts beyond seven days ... since they would deprive the body of this valuable nutrient. It is interesting to note that Paavo Airola, Ph.D., recommends juice fasts whenever a fast is undertaken.

---

[1]Ibid., p. 34.

Right now in the United States there are over ½ **million** people people wearing a colostomy bag. Dr. Walker, as do many others believe **preventive** measures by way of colonics and enemas **far surpass** the *"cure"* of colon removal and the wearing of a colostomy bag the rest of your life!

Cecostomy   *Transverse colostomy*

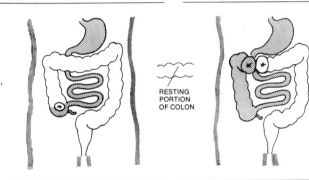

RESTING
PORTION
OF COLON

Descending colostomy   *Sigmoid colostomy*

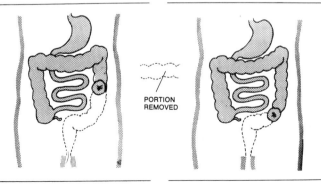

PORTION
REMOVED

14

BRAN . . . THE BOWEL NORMALIZER

High-Fiber Intake Recommended

Donald G. Cooley, along with many medical doctors, reports in his book, <u>After-40 Health & Medical Guide</u>:

> There is little question
> that high-fiber intake
> increases stool weight,
> shortens transit time through the bowels,
> decreases pressures in the colon,
> and produces a stool
> of desirable consistency.
>
> Experience has shown that the
> vast majority of patients with
> diverticula or the irritable bowel
>   syndrome
> improved on high-fiber intake,
> regardless of whether
> their symptoms were predominately
> constipation or diarrhea.
>
> Patients may do well to take
> a cereal bowl full of a bran product
> every day,
> and if constipation is a problem,
> to increase the bran intake
> until the constipation is overcome . . .
>
> It has been a truly educational campaign
> during the past ten years
> to convince dieticians
> that a high-fiber diet
> should be a part of the
> hospital regimen.[1]

[1] Donald G. Cooley, <u>Better Homes and Gardens After-40 Health & Medical Guide</u> (Des Moines, Iowa: Meredith Corporation) 1980, pp. 220, 221.

By restoring the dietary bulk nature intended us to eat, we can help prevent and cure constipation, irritable bowels and other all-too-common ailments.

BRAN HAS MANY ADVANTAGES

Bran
Best

The normal functioning of the intestinal tract depends upon the presence of adequate fiber . . . the kind that absorbs water and forms soft bulk. Bran is useful because it is an undigestible portion of food that passes down, unaltered, into the lower bowel (the colon).

What makes bran so special? Why not just eat a high-fiber diet? I. Taylor, M.D. and H. L. Duthie, M.D., surgeons at the University Surgical Unit at the Royal Infirmary in Sheffield, England found that bran is the one high-fiber product that most effectively *normalizes* bowel function.[1]

Drs. Taylor and Duthie tested a select number of patients with diverticular disease. One-fourth of them received a high-fiber diet; another one-fourth a bulk laxative. The remaining 50% of the patients received bran supplements of about six teaspoons of bran each day. At the end of one month, the treatments were reversed.

All the patients reported some improvement. On the high-fiber diet, 20% experienced complete relief from symptoms such

[1]British Medical Journal, April 24, 1976.

as pain and distension. Those taking a bulk laxative . . . 40% reported similar relief. But of those on the bran regimen fully 60% were free of symptoms!

For Constipation
or Diarrhea

**Bran's
Advantages**

Bran showed its greater effectiveness in several ways:

1. Stool _weight_ was increased to a _greater degree_ than in other high-fiber diets or bulk laxatives.
2. Intestinal transit time . . . a measure of how quickly consumed food moves through the digestive system and is eliminated . . . had its greatest reduction among the bran users.

Bran is not a laxative but, rather, a normalizer of bowel functions. Bran is known to relieve not only constipation but also diarrhea.

Bran _lengthens_ transit times in individuals with chronic _diarrhea._

Bran _shortens_ transit times in individuals with _constipation._

Start with Teaspoonful
at Each Meal

**Daily Use
Of Bran
Suggested**

Denis P. Burkitt, a British doctor suggests that most people should add _two to six tea-spoons_ of unprocessed bran to their diet every day. Because bran flakes are tasteless, most people sprinkle them on breakfast cereal or mix them with fruit juice.

Bran Normalizes Colon Pressure

In those with diverticular disease, muscular contractions to trigger bowel movements cause pressure. In a diseased bowel, this causes the intestinal walls to balloon out of shape intensifying the diverticular problem. Bran, however, restores colon pressure to within normal limits while at the same time, cleanses the colon.

It is usually suggested that bran first be taken in portions of one teaspoon with each of your three meals, gradually working up to two teaspoons each meal (or six a day). Bran can cause flatulence (gas) at first . . . thus the one teaspoon per meal initially. By teaspoon the measurement means a full, rounded teaspoon. If you find you still strain to move your bowels, increase the amount of bran a little until you discover the happy medium. Dr. Neil Stamford Painter, a British surgeon, suggests bran be taken all your life as bran is a part of natural foods and does not have the side effects of laxatives.

Helps Eliminate Hemorrhoids

Bran is also beneficial in eliminating hemorrhoids as bran corrects the problem of constipation. And in so doing, bran helps prevent other diseases that may ensue with clogged bowels.

Helps Avoid Blood Clots

Bran Before Surgery

When anesthesia is administered during an operation, the body becomes limp; breathing is taxed and must be assisted by a machine. The blood gets stagnant, inviting the formation of blood clots.

Dr. Maurice Frohn, consultant surgeon at London's Bethnal Green Hospital makes sure his patients are fed an adequate diet of bran prior to surgery. While other doctors give a drug called heparin to keep the blood flowing, Dr. Frohn uses bran as a natural aid. He states:

> *Bran keeps the bowels moving—*
> *which is imperative after surgery.*
> *Otherwise the colon*
> *becomes overloaded and*
> *causes extreme pressure on the leg veins.*
> *This often leads to clotting*
> *and deep vein thrombosis.*
>
> *Over the past four years,*
> *I've operated on over 1500 patients,*
> *and not one of them has suffered*
> *any post-operative blood-clotting.*[1]

Diseases Can Be Avoided

Captain Thomas L. Cleave, a retired Royal Navy surgeon, shook up the medical community in 1966 by insisting that diabetes, heart disease and obesity, as well as disorders of the digestive tract such as constipation, hemorrhoids and even cancer of

[1]Quentin Van Marle, Bran, Your Legs And Your Life (Emmaus, Pennsylvania: Rodale Press) Prevention, September, 1977, p. 72.

the colon, could all be avoided through a natural, high fiber diet.[1]

Dr. Conrad Latto, an English surgeon, visited a 700-bed hospital in Kashmir, India. The Indians were on a high-fiber diet. In 14 years of operations in that hospital, Dr. Latto reports, there was only one single death from embolism *(the destruction of a vein by a clot).*[2]

CONSTIPATION AND BATHROOM DEATH

Pressure on Veins

Constipation can cause a host of diseases. It can also cause blood clots and what is sometimes called *"bathroom death."*

The Lancet (a British medical journal) of August 15, 1959 states:

> *. . . quite remarkable changes in the size of the veins of the limbs and of the flow through them occur during bathroom straining . . .*
>
> *Early correction of constipation would seem to be a wise precaution in patients with any vascular disease liable to give rise to thrombi (blood clots).*

Straining Inadvisable

Straining while having a bowel movement can bring on angina or the rupture of a blood vessel particularly in those with vascular problems most common with those in the middle years and beyond.

[1]Ibid., p. 72.

[2]Ibid., p. 73.

Constipation, fecal impaction can cause cramps, back pain, frequent urination, abnormal heart rhythms, a sudden rise in blood pressure.

The abnormal stress that constipation puts on the heart is greater than you might imagine. It is not rare for someone to die while in the bathroom.

Nutritionists believe bran is part of the solution to the problem.

Natural BULK Bowel Movers

Unprocessed Bran Most Useful

Bran

We have already discussed bran in detail. Bran is the best type of natural roughage and helps correct the diverticular problems so frequently found in one's latter years.

Bran is not only a source of roughage but it also acts as a bulk-producing agent.

Unprocessed bran is the best form of brand to buy. Processed bran is an ingredient found in many breakfast cereals, bran muffins and all-bran bread and is not as effective as the plain, inexpensive unprocessed variety.

Psyllium A Lubricant

Psyllium

Psyllium is a seed. When soaked in water, these seeds become surrounded with a transparent mucilage, swelling into a gelatinous mass. This soon lubricates the intestines and at the same time stimulates them

**Multiple
Doses
Needed**

to normal activity. They help in forming bulky yet soft stools. Single doses are not effective. They should be taken three times daily over several days together with lots of fluid.

There are commercially prepared bulk-producing laxatives which contain psyllium but these usually also have sugar, sodium bicarbonate and are more expensive. Diabetics should exercise caution in using a commercially prepared formula.

1
ALL ABOUT COLONICS

A <u>**colonic irrigation**</u> involves injecting into the colon a large amount of water, <u>flowing in and out at constant intervals.</u>

This continuous flowing in and flowing out washes out material situated above the defecation area and washes the walls of the colon as high as water can be made to reach. Colonics use 20-30 gallons of water — but only a pint or two at a time. A colon irrigation takes one-half to one hour.

Generally, a series of 6-12 colonics are given to remove the encrustations of the colon and empty out the pockets.

A majority of minerals are assimilated in the colon. If the colon is not functioning at normal levels, the body can become mineral-poor and illness will occur.

Many people have parasites in their colon. It was believed that parasites rob the body of nutrients. The main problem, colonic specialists believe, is the excreta the parasites eject from their body (while in your colon). Such poisons can affect a person's disposition, making him nervous and irritable, unable to cope with each day's problems.

Those specialists who practice colonic irrigation say that colon problems are not simply problems of the elderly. One colonic therapist recalled a case of a 5-year-old boy who was constipated since birth. As a result of his constipation his colon was extended or elongated making for longer transit time. He had a variety of symptoms. He had skin eruptions, was hyperactive and had digestive problems.

There are very few qualified colonic therapists in the United States. They are generally concentrated in high population areas such as Los Angeles, Miami, New York and Chicago.

Depression, nervousness, irritability, frequent crying spells, fatigue and severe constipation are symptoms which suggest a distressed colon. Upon X-ray it is not unusual to find parts of the colon double the size it should be. An area of such a colon can also be filled with parasites. The colon in such an individual can also be elongated and rise high into the diaphragm, causing chest pains. It can mimic a heart attack . . . when gas is present in this colon area. In such cases one colonic treatment is not considered sufficient. Many colonic therapists would recommend 6-10 treatments over a short period of time.

Many people are of the impression that only meat eaters can develop parasites. But even vegetarians can get parasites into their body. Parasites can enter vegetables from the fertilizer that is used in farming. Parasites can enter the skin of someone walking barefoot. There need not be a scratch on the foot. They can enter through the pores in the feet.

Diverticulitis is another problem with which some colonic therapists have good success. Many who have advance cases of diverticulitis believe colonic irrigation is a better first alternative to a colostomy operation. These little sacs (diverticuli) appear on the colon where there is a weak area and accumulate waste material. In colonic therapy, the water washes the affected area.

In **oxygen-colonics.** ... along with water, a carefully regulated amount of oxygen is also introduced into the colon. While the water does its job of washing the colon, the oxygen, therapists say, heals the infected area. Medically speaking, diverticuli are a problem you have to live with the rest of your life. Colonic therapists would disagree. Many believe they can heal this condition.

Colonic therapists seek to cleanse the entire colon with water over a period of several treatments. When successful they can remove fecal encrustations that have lodged in the colon for many, many years. As this putrefied material flushes out, the stench is very strong. It rapidly conveys to the patient how clogged up his system had become.

With a clogged colon often a person will have liver problems. The reason: the liver will pass on its waste into the colon. With a poorly functioning colon, often these poisons cannot get through initially and back up into the liver again.

One colonic therapist reports that 50-60% of the patients he sees have tapeworms and 80% of the patients have some form of parasites. Tapeworms can be 20 feet long in your colon and as thick as your thumb.

Some would argue that colonics *(intestinal irrigations)* do harm by removing normal mucus and by producing colitis. Nutritionists who understand colonic therapy would disagree. Foul, putrefactive masses lodged on colon walls do not contain normal mucus and in themselves would encourage colitis. Therefore, a series of colonics would be deemed highly beneficial to restoring health.

A colonic is not a cure-all. If the individual does not correct his dietary habits the same conditions can recur.

A series of colonic irrigations, therapists suggest, removes practically the whole fecal contents of the colon, softens and removes the large masses of mucus present in mucous colitis and restores the colon to its normal size. Such treatment, they believe, can improve the health conditions of those with arthritis, rheumatism, neuritis and a host of other diseases.

One indication that colon therapy is needed, according to Dr. Cora Smith King is the appearance on the chest, abdomen and back of bright red mole-like spots. These are called little _hemangiomas_ (blood vessel tumors) or _telangiectasia_ (end vessel dilation). Dr. Smith suggests these little hemangiomas are tiny danger flags, signifying intestinal toxemia.[1]

Colonic irrigations were popular many years ago and administered even in hospitals until the late 1950's. Much of the data on colonic therapy was written by medical doctors.

W.A. Bastedo in a Journal of American Medical Association publication reported:

> ... colon irrigations are employed in chronic states of the bowel, such as are encountered in mucous colitis, intestinal putrefactive toxemia and in cases in which a focus of infection is believed to reside in the bowel, as in certain cases of rheumatism, neuritis, secondary anemias ... [2]

Today, colonic irrigations are given by colonic therapists or nurses. They are, as a rule, not available in hospitals but rather in natural health clinics. Some are run in conjunction with chiropractic services. They are generally listed in the yellow pages of the phone books under _Colonic Irrigation._ Check with your doctor.

---

[1]Joseph E.G. Waddington, M.D. Scientific Intestinal Irrigation and Adjuvant Therapy (Chicago: The Bryan Publishers) 1940, p. 28.

[2]W.A. Bastedo, Colon Irrigation (Chicago: Journal of American Medical Association, Council on Physical Therapy) 1932, February 27, 98:734.

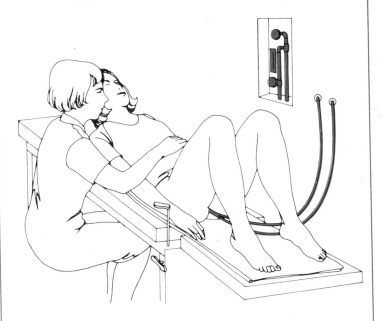

Colonic therapists suggest <u>five</u> ways colonic irrigations are beneficial:

1. Removes practically the whole fecal contents of the colon.

2. Cleanses the delicate mucous membrane of protozoa, bacteria and inflammatory products.

3. Gives valuable information obtainable from the stool concerning physical well-being of patient.

4. Softens and moves the large masses of mucus present in mucous colitis.

5. Removes odors of fermentation and putrefaction.

Colon therapy is a purely physical process of detoxification. Thera-
pists believe, when properly given, the poisonous matter is quickly and completely removed from the <u>entire</u> colon. Many gallons of water are injected into the colon in a steady flow <u>in</u> and <u>out</u> of the colon, thus breaking up the impaction, gently and painlessly.

15

HOW TO DEFEAT DIARRHEA

Maintain Adequate Fluid Intake

The nutritional approach to diarrhea is to try and restore the sufficient presence of friendly bacteria in the colon. It is important that an adequate fluid intake be maintained to replace the water that is lost in the stools . . . thus preventing dehydration.

For adults, some nutritionists recommend daily:

Vitamin A	25,000 I.U.
Vitamin B Complex	200 mg. for 1 week
Niacin	100 mg. three times daily for 2 weeks
Vitamin C	1000-3000 mg.
Magnesium	500 mg.

A generous amount of unflavored, plain yogurt also has been found beneficial to restoring proper bacterial balance in the colon.

Carob And Bananas

Carob flour (found in carob candy as a substitute for chocolate) is rich in a binding substance known as pectin.

Bananas, too, are rich in pectin. They also contain a host of other valued nutrients including, in high quantities: magnesium, potassium phosphorus and Vitamin A.

Dr. Tom Spies, according to a report by Catharyn Elwood, had promising results in treating very small children with apparently incurable diarrhea. He administered niacin therapy and within one hour improvement occurred with all diarrhea gone in 24 hours.[1]

Naturally, diarrhea should be evaluated by your doctor because it can be a symptom of a more complex problem.

The highest sources of niacin are found in the thigh of a stewed chicken (6.2 mg.), beef steak (4 ounces has 8.5 mg.) and brewer's yeast (10 mg.).

Potassium A Balancing Mineral

Dr. Daniel C. Darrow of Yale University has had success treating infant diarrhea with potassium. Potassium is a balancing mineral. Whole grains, black molasses, leafy green vegetables, almonds and potatoes are rich in potassium. Bananas are about the highest source of potassium. One medium banana has 555 mg. of potassium!

For infant diarrhea, nutritionists also recommend cultured milks such as butter-

[1]Catharyn Elwood, Feel Like A Million (New York: Pocket Books) 1976, p. 109.

milk, yogurt, kefir. No raw juices should be taken until condition is corrected. Infant diarrhea can be serious and your doctor should be immediately consulted.

Antibiotics Can Cause Diarrhea

Diarrhea is often the result of antibiotic therapy. Antiobiotics destroy the friendly bacteria along with the unfriendly. Yogurt taken daily or kefir is considered by some an adequate remedy for this.

Kefir is a preparation of curdled milk made by adding kefir grains to milk. It is made by placing 1 tablespoon of kefir grains in a glass of milk. Stir and allow to stand at room temperature overnight. When the milk coagulates, it is ready for eating. Strain and save the grains for the next batch. Kefir is a true *elixir of youth,* used by centenarians in Bulgaria, Russia and Caucasus as an essential part of their daily diet.[1]

[1]Paavo Airola, How To Get Well (Phoenix, Arizona: Health Plus Publishers) 1976, p. 253.

16

HOW TO CONQUER CONSTIPATION

**Bile Acids
Can Be
Hazardous**

Normal to the stool is a high content of bile. With healthy bowel movements, the bile and its chemicals quickly pass through the colon and exit with the stool.

With slow transit time and constipation, the bile elimination is delayed and it becomes subject to the unfriendly action of the abnormal bacteria. It is here that two bile acids which normally are harmless . . . are converted into extremely potent carcinogenic factors.

Researching primitive tribes in Africa, Dr. Denis P. Burkitt found that the high fiber diet of these people speeds the waste matter through the bowels in only 1½ days, while Americans and Europeans living on a highly refined and processed diet require 4 days for food wastes to pass through the bowels. *And therein lies a new definition for constipation.* If it takes more than 1½ days for food wastes to pass through the entire length of your digestive tract, you are constipated![1]

[1]Roland Evin Horvath, Pathways To Living (Hackensack, New Jersey: American Health Education Foundation), Vol. 12, No. 1, p. 2.

TRANSIT TIME . . . THE KEY

**Understanding
Constipation**

Dr. Carlton Fredericks makes an interesting observation:

> *Many laymen assume
> that stool transit time
> is abnormally long
> only when constipation is present,
> but this is a misconception.*

> *One can have a bowel movement
> every day,
> and still be the victim of
> abnormally slow stool transit time—
> because Monday's bowel movement
> should have taken place
> on the preceding Saturday.*

> When constipation
> is added to the problem,
> the situation becomes
> more abnormal and more threatening.

> The threat arises because
> the bowel bacteria in a person
> on a low fiber diet
> are not of the friendly type,
>   which they should be;
> and because,
>   with a delayed stool
>   transit time,
> these bacteria
> vent their unfriendliness
> by breaking down chemicals
>   normal to the stool,
> and converting them into powerful
> carcinogenic (cancer-producing)
>   substances.[1]

---

[1] Dr. Carlton Fredericks, <u>Nutrition Handbook</u> (California: Major Books), 1977, p. 110.

1½ Day
Transit
Time
Ideal

To make it easier to understand . . . transit time means from the time you first eat a meal until the time the residue from the meal is eliminated. Thus, for a 1½ day transit time . . . the Monday evening meal should be eliminated by Wednesday morning. Doctors are able to check transit time from harmless markers that the patient swallows with a meal. This marker appears in the stool.

Some nutritionists believe that normal transit time for a healthy bowel is 12-18 hours, not 1½ days or 36 hours. In one test done by an English medical doctor, a marker pellet was given to a British woman to swallow. This woman moved her bowels every day but did not pass the marker pellet until a week after she swallowed it. You can be "regular" on a daily basis but still be moving food through your insides very slowly.

Many nutritionists suggest that one way to know that toxic cancer-causing chemicals are present in your stool is when the odor is very offensive. As you change to a high residue diet, you will find that the odor of your stool becomes less objectionable.

1
The IRONS Approach To
BOWEL PROBLEMS

V.E. Irons is a well known lecturer in the health field on bowel problems. He graduated from Yale University and heads a company
which produces a number of natural products to improve bowel
function. He is in his eighties.

Mr. Irons, in an interview with The Healthview Newsletter, suggests a
way you can know if your colon is functioning properly:

> When your colon is healthy,
> you will have two well-formed bowel movements
> a day.

> Every morning,
> you'll have a huge movement which should
> altogether be from 2 to 4 feet long.
> Later on in the day,
> you'll have another movement,
> which will be about
> half the size of the first.

> These stools should be expelled effortlessly—
> within seconds after you sit down.[1]

Irons believes there is only one disease . . . " . . . and that disease is
autointoxication—the body poisoning itself. It's the filth in our
system that kills us."

Irons suggests caution on the use of bran. He believes that the best
that bran can do is propel the daily fecal matter through the opening in the center of a clogged colon. Because many colons are so
twisted and clogged and hardened with old feces he does not believe that bran or ordinary food roughage will unplug them. Other
nutritionists see bran as a vital resource to restoring proper bowel
function.

Mr. Irons suggests a 7-day cleansing program to get rid of fecal matter that has built up over the years. For 7 days you eat nothing at all.
Instead he recommends an intestinal cleanser taken 5 times a day
every 3 hours and a volcanic ash substance. The intestinal cleanser
is made from a special grade of psyllium seed. It apparently clings
to the colon walls, holds moisture there, softens and loosens the fecal matter.

---

[1]Sam Biser, The Healthview Newsletter (Charlottesville, Virginia 22906) 1979,
Number 10, page 2.

2
The IRONS Approach To
BOWEL PROBLEMS

The specially compounded volcanic ash acts like a magnetic sponge and Mr. Irons states that it removes toxins from the digestive tract.

For those not up to a 7-day fast, Mr. Irons suggests either a 3-day juice fast every 3 or 4 weeks. He also suggests a food supplement made up of the concentrate of juices of wheat, rye, oat and barley grasses. This comes in tablet form.

Mr. Irons believes that each day one is on a cleansing program that a coffee enema be taken. He suggests that a coffee enema (1) acts as a solvent to help dissolve fecal deposits and (2) stimulates the muscles of the colon. This shakes off the fecal matter that coats them.

Where fecal matter is so encrusted that a coffee enema is ineffective, Mr. Irons recommends an olive oil enema. A gallon of olive oil is needed since it is used full strength.

Mr. Irons suggests that it may take several months or years to repair damage to your colon.

When asked if his treatment was any different from other natural treatments for bowel problems, Mr. Irons replied:

> *It definitely is.*
> *Other natural bowel programs*
> *concentrate on nutrition—*
>   *what you put into your body.*
>
> *My program*
> *emphasizes elimination—*
>   *what you take out of your body.*
>
> *If you have both*
>   *perfect nutrition and perfect elimination,*
> *then you have perfect health.*[1]

A complete interview by V.E. Irons can be secured from The Healthview Newsletter, Box 6670, Charlottesville, Virginia 22906. Request issue Number 10. Single issue is $2.

---

[1] Ibid., p.6.

17

KEYS TO A HEALTHY LIFE

JUNK THE JUNK FOOD DIET

Eliminate Highly Refined Diet

Most Americans eat a highly refined diet that is low residue. This breeds constipation and a slow transit time of 3 or 4 days before it is eliminated. And then, not all is eliminated. Some adheres to the colon walls, disturbing digestive processes and becoming a breeding ground for a host of diseases.

Anyone living on large amounts of refined foods often called the "*soggy bun, hamburger, french fries*" diet will find their intestines becomes cesspools of harmful putrefactive bacteria.

Source Of Odor

It is these putrefactive bacteria that not only cause terrible smelling bowel movements but also produce toxic chemicals which are absorbed into your bloodstream.

Putrefactive bacteria can produce *histamine* which causes many allergies. It is the *histamine* that produces tissue irritants and through its enzymes can destroy B-complex Vitamins. Many people take what is called an *antihistamine* but this only treats the symptom and does not attack the root problem ... which may be an unhealthy bowel!

B-Complex
Essential

Because these putrefactive bacteria destroy B-complex, and because B-complex is essential to peristalsis *(wave-like contractions of the intestines)* . . . many nutritionists suggest a high B-complex supplement for those with these problems. Some suggest a B-complex supplement that contains 50 to 100 milligrams of ALL the B-complex vitamins . . . taken once daily with a meal.

We Are
What
We Eat!

Celiac disease (sprue), Crohn's disease, Colitis, Allergies, Diverticulosis, Hiatal Hernia, Hemorrhoids, Diarrhea, Constipation and Cancer may in one way or another be directly related to unhealthy bowel function and delayed transit time of that which we eat.

DISEASE BEGINS IN THE COLON

The
End Result
Or
The Origin

Those engaged in orthodox medicine tend to treat the end result and not the origin of the problem. Thus a person with cancer of the colon may have the cancerous portion of the colon removed. An individual with Crohn's Disease may be placed on a low residue diet and may have a portion of his intestines removed surgically. An individual with an allergy may find himself getting a battery of allergy tests and eventually taking antihistamines over a long period of time.

The nutritionist is not as concerned with the end result of the problem (or the symp-

A STAGNANT POOL   **A GENTLE, FLOWING STREAM**

Take a mental photograph of your meal each time you sit down to eat. Ask yourself the question: *"Is the food I am about to eat going to help make my body a stagnant pool or will this food make my body a gentle, flowing stream?"* If you do this honestly, chances are your eating habits will change and your overall health will improve.

tom). The nutritionist, instead, seeks the root cause of the problem and by nutrition, vitamin and mineral supplements, exercise, etc. seeks to correct the origin of the problem. There is quite a difference in approach!

It has often been said that America is the land of clogged colons!

Is it any wonder? Recall the food you ate over the past 24 hours. Did it include a greasy hamburger nestled in a soft bun, french fries, coffee and pie? Is it any wonder that many diseases begin in the colon?

How much of the food you ate was as Nature prepared it—raw?

You Can Make The Difference!

A large majority of ill people have one thing in common ... poor bowel habits brought about by poor eating habits.

Your body should become a gentle, flowing stream. If your body resembles a stagnant pool, you are courting illness.

The decision lies with you!

Dr. Paul Eck has a degree in the field of naprapathy. Naprapathy is a system of therapy which attributes all disease to disorders of the nervous system, ligaments and connective tissue.

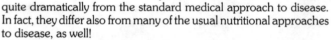

He is director of Analytical Research Laboratories in Phoenix, Arizona which specializes in interpreting hair tests. Paul Eck prepares vitamin and mineral programs for many medical doctors and other health practitioners.

Dr. Eck has very definite views on how to correct basic illnesses. They differ quite dramatically from the standard medical approach to disease. In fact, they differ also from many of the usual nutritional approaches to disease, as well!

Dr. Eck believes that much of what is going on today in the field of nutrition is guesswork. Too many people are spending anywhere from $5000 to $15,000 to correct health deficiences and are no better off physically. The problem, as Eck sees it, is that it is not the products they are taking that are wrong . . . but that the necessary knowledge and application to <u>utilize</u> these various nutrients is missing!

In an interview with Sam and Loren Biser of <u>The Healthview Newsletter</u>, Eck commented:

> *It's not the amount of vitamins that you take nor is it the amount of minerals you are taking or anything else . . . the products or the money that you are spending . . . I think a lot of this is pure waste because if it is not being applied in a scientific way there's no way that it's ever going to work![1]*

---

[1] Paul Eck, <u>Your Minerals and Your Health</u> (To secure this one hour cassette, send $10 <u>direct</u> to Healthview Newsletter, Box 6670, Charlottesville, Virginia 22906)

Other cassettes by Dr. Paul Eck on various metabolic dysfunctions and philosophy of health are available from Analytical Research Laboratories, Inc., 2338 West Royal Palm Road, Suite F, Phoenix, Arizona 85021.

The MINERAL Approach To Illness
An Interview with Paul Eck

When Paul Eck first became interest in hair analysis he discovered he had a low zinc level of 9 (the normal zinc level is 20). He also had hypoglycemia, he was nervous and had a tendency towards diabetes. He also knew his anxiety levels were high.

To try and correct his low zinc level he began taking 3 zinc tablets a day. He was surprised to find that he did not get better. In fact, he became worse and experienced extreme fatigue. He decided to go off the zinc and his energy level began to come back to what it was before.

Simply adding zinc to your system can cause other problems. Too much zinc has a tendency to make your copper level low. The normal level for copper in the human body, according to Eck, is 2.5mg./% He states that anyone with a copper level of 1.0 is in a potential cancerous state.

Low copper levels symptoms may show up as: fatigue, anemia, joint pain and depression. Eck believes this can lead to a cancerous condition because:

> ... the copper inside the cell regulates the respiratory oxidation mechanism which prevents you from getting cancer.[1]

Paul Eck also says that if one has a high copper level ... and takes a multiple vitamin/mineral tablet or supplement that has copper in it, their physical problem will become worse. The additional copper, adding to their already high level, will accentuate their illness and the other vitamins and minerals in the daily supplement will be of no value. The indiscriminate taking of vitamins and minerals can be a hit and miss type of therapy. It can be a total waste of money, according to Eck. He believes that up to 90% of the people taking multiple vitamin supplements are damaging their health.

Eck states that copper is the most dangerous of all the required minerals, if you are not aware of what your body requires. Simply to take copper supplements on a guess-basis can be hazardous to your health.

---

[1]Ibid.

Paul Eck also believes that excessive Vitamin C taken over a considerable length of time and in certain "biological types" can cause cancer. He suggests that soils that are high in calcium and magnesium protect against cancer. Any mineral, therefore, that can lower these two minerals (calcium and magnesium) or copper can cause a cancerous condition.

In a recorded interview for <u>The Healthview Newsletter,</u> Eck stated:

> *There is no question that a person can take Vitamin C in some cancer cases and derive results.*
>
> *But, just as zinc lowers copper, which is perhaps the most protective mineral against cancer ... Vitamin C also has a copper lowering effect, and over the years, can cause cancer.*[1]

Dr. Paul Eck does not believe it is necessary in the majority of cases to give large doses of Vitamin C. He says that copper makes up a part of an enzyme called <u>ascorbic acid oxidase.</u> An <u>oxidase</u> is a catalyst. It is the copper in this formula that activates the enzyme. If this ascorbic acid oxidase enzyme is missing, Eck believes then that Vitamin C cannot be oxidized to adequate amounts in the body. Therefore, Eck concludes, if this optimal oxidation is not taking place, large amounts of Vitamin C are not only useless but can initiate various disease proccesses.

If your body is not functioning properly, the vitamins will not excrete the excess minerals you may be taking in your daily supplements. Instead they accumulate in the body. If your body is deficient in Vitamin B[6], as an example, you will have a tendency to accumulate copper in the body. This can result in a <u>toxic</u> level of <u>copper</u> in your system.

Dr. Eck does not encourage the taking of multiple vitamins and minerals. He feels this indiscrimate use is detrimental to one's health ... since every individual is biochemically different.

---

[1]Ibid.

Paul Eck is a firm believer in using hair analysis tests. The reason, he states:

> A hair analysis test is the only method developed that has any validity at all as far as measuring what actually is occurring in the tissues of the body.
>
> It has the benefit of being able to give you a metabolic pattern of many metabolic activities that are occurring in your body over a period of time.[1]

Dr. Eck believes that blood tests, in this context, are frequently invalid because they give you an up-to-the-minute readout. It is not a true reflection of what is happening in the tissues over a period of time. A person taking a high amount of Vitamin C could be releasing from his system large amounts of cholesterol. If a blood test were taken at that time it would show a high cholesterol level. However, what the physician does not realize is that it is not a build-up of cholesterol but, quite the opposite, a beneficial flushing of cholesterol out of the body.

Too many people, Dr. Eck suggests, take manganese when they have a manganese deficiency . . . they take iron when they have an iron deficiency, etc. **This is wrong.** To give iron to raise iron is to lower iron!

Constipation And Its Causes

Dr. Eck states that the major causes of constipation are an **(1)** underactive thyroid gland **or** an **(2)** underactive adrenal gland **or (3)** both.

The optimal THYROID RATIO is a calcium/potassium ratio of 4 parts of calcium to one part of potassium. When the calcium/potassium ratio is chronically greater than 10/1 with a concurrent constipation, an underactive thyroid gland is responsible.

The optimal ADRENAL RATIO is a sodium/magnesium ratio of 4.16 parts of sodium to one part of magnesium. When the sodium/magnesium ratio is chronically less than 2/1 with a concurrent constipation, an underactive adrenal gland or adrenal insufficiency is responsible.

When you go on a correct vitamin/ mineral supplementation, the excess minerals and toxic minerals (such as cadmium, lead, aluminum) will start to unload and flush out of your system. This unloading will cause headaches and numerous other symptoms which will vary with the toxic metal or combination of toxic metals being eliminated, and in some cases make you feel worse. This is a natural occurrence as your body gets rid of these unwanted elements to get you on the road to full recovery.

Dr. Eck believes that mineral imbalances should be corrected mainly by using <u>small</u> potency vitamins and minerals. He says it just takes a very small amount of a mineral to initiate a major physiological process in the body. Any amount over that, he states, will cause exactly the opposite reaction.

Paul Eck believes that mineral therapy is also indirectly hormone therapy. Through the results found in hair analysis he has been able to see people go off hormone therapy, estrogen, even off of thyroxin. He has used manganese and copper to improve their thyroid function where indicated.

Paul Eck is very familiar with Diabetes. His grandmother, his mother was diabetic. He and his brother were pre-diabetic. Dr. Eck states he can determine from a hair analysis, years in advance, whether a person will become a diabetic.

Paul Eck says that 90% of the people who have diabetes have more than enough insulin circulating in their blood. When you have a low calcium to magnesium ratio (such as 3.3 to 1) you have an individual that has diabetes. And if the ratio is high, such as <u>10</u> parts calcium to <u>1</u> part magnesium . . . you are in the diabetic area.[1]

Eck states:

> *One problem in about 10% of the diabetics is the lack of calcium in the pancreas. This condition results in an inability of the Islets of Langerhans to secrete insulin . . .*

---

[1] The calcium to magnesium <u>ratio</u> normally is 6.7 to 1. This means for every 1 part of magnesium in your system, you should have 6.7 parts of calcium.

because when the calcium drops below a certain level you can't even initiate the secretion of the insulin that is manufactured and is being stored in pancreatic Islets of Langerhans tissue.

So what you have to do is, by one means or another, raise the calcium level back up to a close to normal ratio between the magnesium and then you automatically get a secretion of insulin.

This occurs in your insulin-deficiency diabetics . . . which only accounts for about 10 or 12% of the cases.

Dr. Eck states that the rest of the problem in diabetes lies either in the transport of the insulin to the cell itself or, when it gets to the cell, there is a lack of a receptor at the cell site on the cell membrane. Those receptors are all minerals! Therefore, Dr. Eck concludes:

If the proper mineral is not available in the body for transport of the insulin to the cell . . . it doesn't get there in the first place.

Secondly, even when it gets there, if some receptor is not present, and there are multiple receptors on the cell membrane . . . then, of course, the insulin can't even enter the cell and do what it is supposed to do!

We have had such great reports especially in diabetes.

You know the old saying that says "Once you've been on insulin, you're going to stay on it the rest of your life . . ." that's the same statement they use for hypothyroidism. They say: "Once you're on thyroid, you're going to be on it forever. Make up your mind to it."

Some individuals have been able to have their insulin requirement reduced or completely eliminated within a few weeks. These, of course, are spectacular cases. There are also insulin-taking diabetics who require a year or two to bring about a complete correction. The individual must be extremely cooperative. Attempts to correct diabetes must be done under the supervision of a doctor.

Dr. Eck says that those taking oral hypoglycemic agents for diabetics are the easiest cases to correct. Correcting those with juvenile diabetes is much more difficult . . . unless the individual faithfully stays on the health program.

If the ratio remains chronic . . . Dr. Eck believes there are 7 mineral clues as to whether a person is developing cancer. The more of these clues they have the more severe their condition is. Here are the clues:

* 1. Calcium/Magnesium ratio of less than 2 parts of calcium to 1 part of magnesium . . . is a cancer indicator.

* 2. Calcium/Magnesium ratio of over 14 parts of calcium to 1 part of magnesium . . . is a cancer indicator.

* 3. Sodium/Potassium inversion. Normally sodium is 25 in ratio to your potassium, which is 10. This is a 2.5 to 1 sodium to potassium ratio. If the ratio inverts (goes lower than 1.5 to 1) this could be indicative of cancer, kidney disease, hypertension, infections, osteoarthritis, etc.!

* 4. Zinc/copper ratio of over 16 parts of zinc to 1 part of copper . . . is a cancer indicator.

* 5. Zinc/copper ratio of less than 4 parts of zinc to 1 part of copper . . . is a cancer indicator.

* 6. Copper greater than 10 and less than 1.0 irregardless of ratios . . . is a cancer indicator.

* 7. Iron greater than 10 and less than 1.0 . . . is a cancer indicator.

Paul Eck believes that there are important interrelationships between minerals and vitamins. An excess of one mineral can cause an imbalance in another mineral in your body. Such imbalances can lead to illness. Here are some examples:

1. MANGANESE

Manganese can lower magnesium levels in the body if your magnesium level is already low, the additional lowering by taking manganese can cause epileptic seizures and other neuro-muscular dysfunctions.

2. CALCIUM

Whenever you take large amounts of calcium, Eck states you will lose potassium. He says about 80% of the people in the United States suffer from a sluggish thyroid. This causes a high blood cholesterol, lack of incentive, fatigue. Eck says:

---

*These ratios figures are Paul Eck's testing figures. They are not standard ratio figures. What other testing laboratories for hair analysis may consider a normal ratio . . . Eck may consider not in the normal range.

Potassium is necessary for thyroxin, which is a hormone of the thyroid gland.

Therefore, if one takes calcium causing a lowering in potassium he will have a lowering of thryoid function.

Calcium will also drive magnesium out of the body causing a high level of phosphorus to occur and make one prone to dental cavities.

3. Vitamin B_1

Large amounts of Vitamin B_1 can over a period of time cause a manganese deficiency. Initially, the taking of Vitamin B_1 *(thiamine)* will give you a burst of energy. The excess of this B vitamin may also cause a magnesium deficiency. Both manganese and magnesium are important, Eck says, in blood sugar problems. Because of this manganese/magnesium dificiency, Eck believes, they can develop over 70 different diseases . . . including diabetes.

4. IRON

Iron supplements can cause a copper deficiency. When too much iron is taken, Eck states, you can cause extremely high blood pressure, migraine headaches, and arthritis. Many arthritics have iron deposits in the joints of the body. Eck also reveals:

Over 51% of all the cases of heart disease have been found to have iron pigment deposits in the cardiac cells of the heart . . . largely from taking too much iron or an inability to properly metabolize iron.[1]

To give iron to raise iron is to lower iron. This is true of every mineral. When you have an iron deficiency, you give manganese.

5. ZINC

Zinc supplements can cause a copper deficiency resulting in a severe anemia. By causing a copper deficiency the following conditions may result—menstrual problems, prostrate disorders, allergies, arthritis and insomnia to name a few.

6. COPPER supplements can over a period of time result in a Vitamin C deficiency. Excessive copper can also cause a Vitamin B-1 and B-6 deficiency.

---

[1]Ibid.

Impotency and frigidity problems are intimately associated with mineral ratio imbalances caused by "stress," diabetes, hypothyroidism, adrenal insufficiency, etc.

The **seven** main indicators, from a hair analysis, of impotence in a male or frigidity in the female are:

1. A <u>3.3</u> to <u>1</u> or less of <u>calcium to magnesium</u> level indicates that the individual has sexual problems of impotence or frigidity. This inverted ratio is found particularly in diabetics.

2. <u>Sodium/Potassium</u> inversion. The normal ratio is 2.5 of sodium to 1 of potassium. If this is inverted (less than 1.8/1), it is an indicator of sexual problems.

3. <u>Copper.</u> A very high copper level is another indicator.

4. <u>Zinc.</u> Extremely low or high zincs can also indicate sexual dysfunction and be a cause of impotence or frigidity.

5. A Sodium/Magnesium ratio greater than   18/1.

6. A Sodium/Zinc ratio greater than   8/1.

7. A Calcium/Sodium ratio greater than   10/1.

In the problems of obesity (overweight), Dr. Eck breaks down individuals into broad categories such as:

> Fast oxidation
> Slow oxidation
> Mixed oxidation

There are   3   different ways people metabolize their food. Dr. Eck refers to this as <u>Oxidation.</u> Oxidation is the use or burning of foods to produce energy on a cellular level. Regarding oxidation, Dr. Eck identifies the categories and suggests:

1. They are so <u>fast</u>
they are breaking down their sugars very rapidly and they have a great increase in heat production as a result. They are the type of people who, when they eat, they perspire a lot. They are <u>*"fast oxidizers."*</u> A <u>*"fast oxidizer"*</u> is a person who has a <u>hyper</u>active thyroid and <u>hyper</u>active adrenal glands. They tend to have excessive energy levels due to the fast burning of foods, followed by exhaustion.

2. They are so <u>slow</u>
that they are metabolizing their foods very slowly. They are
"slow oxidizers." They have a <u>hypo</u>active thyroid and <u>hypo</u>active
adrenal glands. Slow oxidizer's energy levels are usually low.
This can be due to a number of factors such as the body's
inability to completely break down the foods consumed when a
HCl (Hydrochloric acid) deficiency is present. Low thyroid and
adrenal activity also contributes to slow oxidation as well as toxic
metal accumulation and dietary habits.

3. They are <u>mixed</u> oxidizers
who may be fast in one glandular area and slow in another.
They tend to have energy swings as well as mood swings. This is
due to the *"seesaw"* effect from fluctuating into fast and slow
oxidation.

Both the fast and slow oxidizers are handling their foods the wrong
way.

The Pill Destroys Sex Life Of Women

Dr. Eck believes that the Pill has destroyed the sex life of at least 10
million women.

The Pill creates a false pregnancy. The taking of a birth control pill
raises the copper levels in the body. This creates a mineral imbalance
which lowers your thyroid function as well as adrenal activity.

When a person has a low thyroid activity (<u>hypo</u>thyroid), they don't
have anywhere near the sex arousal . . . nor do they have a strong sex
desire. They don't have the energy for it! Not only that, but the Pill
brings with it menstrual period irregularities and menopausal disorders.

The male with <u>high</u> copper levels also develops a **slow** sexual
arousal. Food that are high in copper include Brazil nuts, peanuts,
sesame seeds, corn grits, broiled cod, baked flounder, broiled halibut,
steamed lobster, pike and perch, ham, liver. <u>Oysters are extremely
high in copper.</u> Just 1 cup of oysters (cooked, fried or raw) contains
59 milligrams of copper!

You also take copper into your body by drinking water coming through copper water pipes or cooking out of copper cookware.

Some women use a copper IUD birth control device. Because vaginal secretions are normally acid . . . that acidity leaches the copper off the coil and it goes into your system. Dr. Eck states that:

> *It is estimated that there is enough copper*
> *eroded from a coil in one year*
> *to actually cause susceptible individuals*
> *to become schizophrenic.[1]*

In a pregnant woman, the copper keeps building up during pregnancy. The fetus stores a large amount of copper that he gets from his mother's liver. This usually lasts the child for 12 years.

> *At the end of 12 years . . . if the child's copper level*
> *does not go down . . . you have females complaining of*
> *acne and adolescent problems, etc.*

If the mother cannot quickly unload the copper excess after pregnancy . . . she develops postpartum depression. Some women have become mentally unbalanced after giving birth. Paul Eck believes that copper excesses are the problem. Depending on the mineral imbalances of the individual, Eck uses either minerals or vitamins to unload the copper excesses. The hair analysis determines what supplements are needed.

What Results Can One Expect

Dr. Eck states that those who have a hair analysis and follow through on a personalized supplement program will experience symptomatic changes within two to three weeks. At least one year on supplements is needed to approach normalized mineral levels. Toxic metals and toxic minerals can be eliminated from the tissues in about 6 months.

They may experience periodic worsening of their general condition depending upon their findings. As an example, if a person has rheumatoid arthritis, many times within the first two weeks there may be a marked reduction in pain. However, if the individual has

---

[1]Schizophrenia is *a major mental disorder typically characterized by a separation between the thought processes and the emotions . . . a distortion of reality, accompanied by delusions and hallucinations.*

numerous heavy metal accumulations, the removal from tissues and joints of these toxic metals will trigger a temporary flare-up in their condition. This may occur several times throughout the program. Dr. Eck suggests that if this flare-up of symptoms becomes too severe, the individual should reduce or stop taking his supplements for a few days until the symptoms subside.

Permanents, tints, bleaching and coloring of hair does not make any insurmountable significant changes in hair analysis mineral readings. Some shampoos, hair treatments do affect mineral levels, however.

Selsun Blue may cause an elevation in selenium levels.
Head and Shoulders or Breck may result in elevated zinc.
Grecian Formula or other darkening agents will many
    times result in elevated lead levels. Lead acetate is used
    in these products to blacken the hair.

Dr. Eck suggests that hair analysis retests should be done three months after the first test to check progress. If the individual is a *"mixed oxidizer"* or *"fast oxidizer,"* a retest is suggested in two months.

Dr. Eck says you cannot treat these people exactly the same as far as anything is concerned. You must take into consideration a broad classification of their oxidation types . . . preparing a program on that premise. You cannot give any one mineral for an obesity problem or any other problem. Hair analysis will determine what minerals are deficient and what minerals are in excess.

Paul Eck is very sold on proper hair analysis. In fact he is so sold on the necessity for hair analysis that he would not suggest any mode of treatment for any condition . . . to a physician . . . until a hair analysis of the patient has been made.

Hair analysis is becoming more and more popular. And there are quite a few hair analysis laboratories throughout the United States. Not all agree with Dr. Paul Eck's approach. In fact, he may be considered a maverick in the field. But his laboratory in Phoenix is kept very busy. It could be a sign that his customers are getting excellent results from his recommendations.[1]

---

[1]Dr. Paul Eck, Analytical Research Labs, Inc., 2338 West Royal Palm Road, Suite F, Phoenix, Arizona 85021.

Bibliography
Recommended Reading

Airola, Paavo, *How To Get Well*, Health Plus Publishers, Phoenix, Arizona, 1976.

Biser, Sam, *Healthview Newsletter*, Nos. 10, 11, Charlottesville, Virginia 22901.

Cooley, Donald G., Better Homes and Gardens *After-40 Health and Medical Guide*, Meredith Corporation, Des Moines, Iowa, 1980.

Elwood, Catharyn, *Feel Like A Million!*, Pocket Books, New York, 1976.

Moyle, Alan, *Everything You Want To Know About Diets To Help Constipation*, Pyramid Books, New York, 1972.

Nourse, Alan E., M.D., *Ladies' Home Journal Family Medical Guide*, Harper & Row, New York, 1973.

Rothenberg, Robert E., M.D., *The Complete Surgical Guide*, Weathervane Books, New York, 1974.

Smith, William, *Herbs For Constipation*, Thorsons Publishers Limited, Northamptonshire, England, 1976.

Tilden, John H., *Toxemia*, The Basic Cause Of Disease, Natural Hygiene Press, Chicago, Illinois, 1974.

Troy, Marian T., *Better Bowel Health*, Pyramid Books, New York, 1974.

Warmbrand, Max, *The Encyclopedia of Health and Nutrition*, Pyramid Books, New York, 1974.

FOR YOUR <u>LIFE</u> ... KEEP INFORMED!

Now available! A quarterly Total Health Guide Newsletter to keep you up-to-date on the <u>very latest</u> of Medical/Nutritional Data! You owe it to yourself and to your loved ones ... to be fully informed! Now! At last! You can have the <u>most current</u> information on diseases and their treatment ... even before it is available to the general public! The information you receive may help save your life ... or the life of a loved one!

<u>TWO</u> WAYS TO SUBSCRIBE

1. Total Health Guide Newsletter

Quarterly, for one year we will send you the 8-page Newsletter containing all the latest data on the major diseases. The Newsletter will present an unbiased report on both the Medical and Nutritional discoveries plus reports on their effectiveness and availability.

One Year: $25

2. Total Health Guide Newsletter
Plus
PERSONALIZED TYPEWRITTEN UPDATE

You will receive the quarterly 8-page Newsletter which reports the latest in Medical and Nutritional approaches to disease.

Plus! You will also receive a <u>TYPEWRITTEN REPORT</u> on the <u>specific disease in which you are personally interested.</u> This TYPEWRITTEN REPORT will be mailed to you within 3 weeks after you subscribe.

- You will also receive a <u>RING BINDER</u> to hold the Newsletters and Report. It will also contain a special unit to hold cassettes.

- Plus you will be sent the <u>cassette</u> ...
  BALANCING YOUR EMOTIONS
  by Dr. Jonas Miller.

One Year: $50

SEE
OTHER
SIDE

SALEM KIRBAN, Inc., Kent Road, Huntingdon Valley, Pennsylvania 19006

- -

YES! I want to keep informed! Send me the Health Information Service I have checked below. My check is enclosed.

☐ 1 Year / $25
Total Health Newsletter

☐ 1 Year / $50 *(Fill in other side)*
Total Health Newsletter
Typewritten Health Report
(Includes Ring Binder/Cassette)

Mr./Mrs./Miss   (Please PRINT)

Address

City   State   ZIP

REQUEST For Personalized TYPEWRITTEN HEALTH UPDATE

If you are subscribing for Medical/Nutritional Health Information for One Year at $50, you are entitled to a Typewritten Health Update on one disease.

It is important you understand that we neither diagnose or prescribe. Therefore we **cannot** make personal recommendations to you. Only your physician can do this. What we do provide you is the very latest in both Medical and Nutritional information on the disease in which you are interested.

Each month we go through hundreds of publications, books and listen to both medical and nutritional seminar cassettes. We cull from all of this the data that is essential to the particular disease in which you are interested.

We would be happy to answer specific questions in this TYPEWRITTEN HEALTH UPDATE providing they are not in the realm of diagnosing or prescribing.

DISEASE I want Data on_____

MY QUESTIONS I particularly would like answered: *(Please PRINT)*

1. _____

_____

_____

2. _____

_____

_____

3 _____

_____

_____

4. _____

_____

_____

**FILL IN
RESPONSE FORM
ON
REVERSE SIDE
AND
MAIL WITH YOUR CHECK**

Use this ORDER FORM to order additional copies of

UNLOCKING
YOUR BOWELS
FOR
BETTER HEALTH
by Salem Kirban

Your loved ones and friends will find
this book invaluable! Why not give this
excellent book to those who want hon-
est answers to their problems.

QUANTITY PRICES

1 copy: $4.95

3 copies: $12 (You save $2.85)
5 copies: $20 (You save $4.74)

ORDER FORM

Salem Kirban, Inc.
Kent Road
Huntingdon Valley, Pennsylvania 19006

Enclosed find $ _____ (plus $1 postage) for _____ copies of

UNLOCKING YOUR BOWELS FOR BETTER HEALTH
by Salem Kirban

Name_____
    Mr./Mrs./Miss   (Please PRINT)

Street_____

City_____

State_____Zip Code_____

NOW! ... You can make intelligent, life-changing decisions when you know **both approaches** to correcting ailments that plague you or your loved ones!

THE MEDICAL APPROACH
versus
THE NUTRITIONAL APPROACH

NEVER BEFORE ... in one book ... has an unbiased comparison been outlined, clearly, simply, showing both the Medical approach versus the Nutritional approach to major diseases!

At last! Sixteen books are now available! Each book defines the disease in words you can understand plus graphic pictures. The symptoms are also outlined.

Each book shows how medical doctors approach the examination of the patient, what tests they conduct, what drugs they recommend (and their side effects), the type of surgery followed and what their prediction is regarding the course of the disease and the probability of recovery (termed, *prognosis*).

In the same book, you will also read the Nutritional approach to the same disease; what natural therapy has been used through the years, what results have been achieved and what the prognosis is using nature's way.

Save by buying several books. Give to loved ones. You may be giving a gift of LIFE! **Each book is $5.**

The Medical Approach versus The Nutritional Approach

ARTHRITIS
by Salem Kirban ①

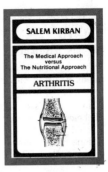

What causes arthritis? How many types of arthritis are there? Does medicine help or hinder? Is chiropractic treatment valid? How will the disease progress if not corrected? What do physicians recommend? What is the nutritional approach to the same problem?

How to recognize the symptoms of arthritis. What diet do many nutritionists feel is beneficial? What foods should I avoid? Answers to these and more!

The Medical Approach versus The Nutritional Approach

CANCER
(including
Breast and Lung) ②
by Salem Kirban

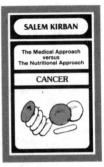

What causes cancer? What do nutritionists believe is the cause of cancer? What are the basic types of cancer? Is surgery the answer? What about chemotherapy? Can drugs cure cancer? What side effects can I expect?

Is a proper nutrition program effective against cancer? What foods should I eat? Should I go on a juice diet?

The Medical Approach versus The Nutritional Approach

HEART DISEASE
③
by Salem Kirban

Does a diagnosis of heart trouble mean the end is near? Can I do something about it and live a happy, healthy long life . . . even after a heart attack?

What about drugs and Vitamin E? What is the sensible nutritional approach to the problem? How can I regain a sense of well-being and abundant energy without fear? What foods should I avoid? How can I flush my system clean again?

The Medical Approach versus The Nutritional Approach

HIGH BLOOD PRESSURE
④
by Salem Kirban

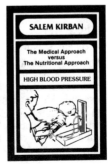

Why is high blood pressure dangerous? What are the causes? Is there any way nutritionally to lower my blood pressure? What drugs do medical doctors prescribe? What are the side effects? Do these "miracle" drugs really work?

What is the nutritional approach to high blood pressure? What juices should I drink? What vitamins and minerals are beneficial? Is fasting beneficial? What foods should I eat?

The Medical Approach versus The Nutritional Approach

DIABETES
by Salem Kirban

⑤

What causes diabetes? Must I change my lifestyle? Why do medical doctors prescribe insulin? What is the prognosis for one who is told he has diabetes?

Can a supervised nutrition program minimize the effect of diabetes? Will it provide a normal lifestyle? What foods should you eat? What juices are beneficial? Does the water I drink make a difference? Are vitamins and minerals and herbs worthwhile?

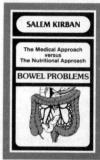

The Medical Approach versus The Nutritional Approach

BOWEL PROBLEMS
by Salem Kirban

⑥

How can I unlock my bowels for better health? How can I achieve that vibrant vitality again and gain that schoolgirl complexion? How can I break the laxative habit? Are drugs the answer?

How can I get rid of hemorrhoids forever? What vitamins and juices are especially beneficial? Are suppositories worthwhile? If so, what type? Can you have daily bowel movements and still be constipated?

The Medical Approach versus The Nutritional Approach

PROSTATE PROBLEMS
by Salem Kirban

⑦

What are the early warning signs of prostate problems? What drugs do medical doctors recommend? What are the side effects? What surgery do they recommend? Is the cure worse than the problem?

Does the nutritional approach offer a more lasting alternative? What diet is recommended? How can you avoid prostate problems in sexual union? Why waiting to correct the problem is dangerous! Do juices and vitamins help?

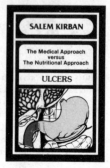

The Medical Approach versus The Nutritional Approach

ULCERS
by Salem Kirban

⑧

What causes gastric and duodenal ulcers? Are the "miracle" drugs really effective or do they bring with them a host of insidious side effects? What warning signals give you advance notice of an impending ulcer?

What foods are especially helpful? Are juices beneficial? Which ones and how should they be taken? What may happen if you don't change your way of life? What vitamins, minerals are beneficial?

The Medical Approach versus The Nutritional Approach

KIDNEY DISEASE ⑨
by Salem Kirban

What is the medical approach to kidney disease? What are some of the problems that can develop if the disease is not nipped in the bud? What are the side effects of the drugs prescribed?

Is meat harmful? What type of diet is beneficial? Is a supervised fast recommended? How long? What common, ordinary foods and juices have proven beneficial? What vitamins and minerals help? What about herbs?

The Medical Approach versus The Nutritional Approach

EYESIGHT ⑩
by Salem Kirban

Are glasses the answer to failing eyesight? Is the medical approach to Cataracts the only solution? What do nutritionists recommend?

What eye exercises may prove beneficial for my eyes? Can I throw away my glasses? Is poor eyesight an indication of other growing physical problems? Can diet correct my poor eyesight? What juices may prove beneficial? What combination of vitamins and minerals should I take?

The Medical Approach versus The Nutritional Approach

IMPOTENCE/ FRIGIDITY ⑪
by Salem Kirban

Impotence is the incapacity of the male to have sexual union. Frigidity is the incapacity of the female for sexual response. Both of these problems are growing because of today's stressful lifestyle! They lead to other trials!

What is the medical approach to these problems? How successful are they? What is the nutritional approach? What type of diet is recommended? Do juices help? Are herbs beneficial? Much more!

The Medical Approach versus The Nutritional Approach

COLITIS/CROHN'S DISEASE ⑫
by Salem Kirban

What causes Colitis? What drugs do doctors recommend? What are the side effects: How successful is surgery? What is the nutritional approach to Colitis? What foods are beneficial? What about juices and vitamins?

What is Crohn's Disease? What are the symptoms? Why does it recur? What is the medical approach to the problem? What is the nutritional approach? Can juices and vitamins correct the cause?

HOW TO BE YOUR AGAIN
(13)

by Salem Kirban

How can I restore my energy and eliminate fatigue? How can I develop Reserve Energy as an insurance to good health and a hedge against illness? How can I begin a simple, day by day health program?

How much should I eat and when should I eat? How can I check my own Nutrition Profile daily? How can I feel like 20 at age 60? How can I turn my marriage into a honeymoon again? When should I take vitamins, minerals? What juices are vital for a youthful life?

OBESITY
(14)

by Salem Kirban

What causes Obesity? Why don't fad diets work? Is being overweight a glandular problem or a dietary problem? Is obesity a liver, pancreas or thyroid problem?

What is the medical approach to treating those who are overweight? What illnesses will obesity encourage? Why does the nutritionist treat your colon? What nutritional approach will take off weight easily and permanently giving you a new lease on life?

HEADACHES
(15)

by Salem Kirban

What causes headaches? Can headaches cause depression and hypoglycemia? Are women more prone to have nagging headaches? Must you live with migraine headaches all your life? What is the medical approach to headache problems?

Can a proper nutritional approach rid you of migraine headaches? Do vitamins help? Is fasting beneficial.? What about the pressure point techniques? What 3 herbs promise headache relief? How to tell your migraine *"Goodbye!"*

HYPOGLYCEMIA
(16)

by Salem Kirban

Is Hypoglycemia a fact or a fad? Why has this word . . . Hypoglycemia . . . become the focus of intense controversy? Is it the cause of many unexplained ills? What is the medical approach to this problem?

Anxiety, irritability, exhaustion, lack of sex drive, constant worrying, headaches, indecisiveness, insomnia, crying spells and forgetfulness . . . are these all signs of Hypoglycemia? How do nutritionists approach this problem with diet and supplements? Will this approach give you a new life?

Quantity	Description	Price	Total

The MEDICAL APPROACH Versus The NUTRITIONAL APPROACH Series

Quantity	Description	Price	Total
_____	1 Arthritis	$ 5.00	_____
_____	2 Cancer	5.00	_____
_____	3 Heart Disease	5.00	_____
_____	4 High Blood Pressure	5.00	_____
_____	5 Diabetes	5.00	_____
_____	6 Bowel Problems	5.00	_____
_____	7 Prostate Problems	5.00	_____
_____	8 Ulcers	5.00	_____
_____	9 Kidney Disease	5.00	_____
_____	10 Eyesight	5.00	_____
_____	11 Impotence and Frigidity	5.00	_____
_____	12 Colitis/Crohn's Disease	5.00	_____
_____	13 How To Be Young Again	5.00	_____
_____	14 Obesity	5.00	_____
_____	15 Headaches	5.00	_____
_____	16 Hypoglycemia	5.00	_____
_____	**All 16 Health Books** *(Save $30)*	**$50.00**	_____

Single Book	$5	All 50 Books	$50*
Any 3 Books	$12	(*You save $30)	

Other SALEM KIRBAN HEALTH BOOKS

Quantity	Description	Price	Total
_____	Unlocking Your Bowels For Better Health	4.95	_____
_____	How Juices Restore Health Naturally	4.95	_____
_____	How To Eat Your Way Back To Vibrant Health	4.95	_____
_____	How To Keep Healthy & Happy By Fasting	4.95	_____
_____	The Getting Back To Nature Diet	4.95	_____
_____	**How To Win Over IMPOTENCE/FRIGIDITY**	6.95	_____
	(Expanded Version with Full Color Section)		_____

Total for Books _____
Shipping & Handling _____
Total Enclosed $ _____

(We do NOT invoice. Check must accompany order, please.)

When using Credit Card, show number in space below.

☐ Check enclosed

☐ Master Charge

☐ VISA

When Using MasterCard
Also Give Interbank
No. (Just above your
name on card)

Card Ex-pires	Month	Year

***POSTAGE & HANDLING** Use the easy chart to figure postage, shipping and handling charges. Send correct amount and avoid delay.

TOTAL FOR BOOKS	Up to 5.00	5.01-10.00	10.01-20.00	20.01-35.00	Over 35.00
DELIVERY CHARGE	1.50	2.00	2.50	2.95	NO CHARGE

FOR ADDITIONAL SAVINGS: Orders Over $35.00 Are Now Postage-Free!

SHIP TO _____

Mr./Mrs./Miss   (Please PRINT)

Address _____

City _____ State _____ ZIP _____

SALEM KIRBAN, Inc./Kent Road, Huntingdon Valley, Pennsylvania 19006

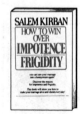

Quantity	Description	Price	Total

The MEDICAL APPROACH Versus The NUTRITIONAL APPROACH Series

Quantity	Description	Price	Total
_____	1 Arthritis	$ 5.00	_____
_____	2 Cancer	5.00	_____
_____	3 Heart Disease	5.00	_____
_____	4 High Blood Pressure	5.00	_____
_____	5 Diabetes	5.00	_____
_____	6 Bowel Problems	5.00	_____
_____	7 Prostate Problems	5.00	_____
_____	8 Ulcers	5.00	_____
_____	9 Kidney Disease	5.00	_____
_____	10 Eyesight	5.00	_____
_____	11 Impotence and Frigidity	5.00	_____
_____	12 Colitis/Crohn's Disease	5.00	_____
_____	13 How To Be Young Again	5.00	_____
_____	14 Obesity	5.00	_____
_____	15 Headaches	5.00	_____
_____	16 Hypoglycemia	5.00	_____
_____	**All 16 Health Books *(Save $30)***	**$50.00**	_____

Single Book	$5	All 50 Books	$50*
Any 3 Books	$12	(*You save $30)	

Other SALEM KIRBAN HEALTH BOOKS

Quantity	Description	Price	Total
_____	Unlocking Your Bowels For Better Health	4.95	_____
_____	How Juices Restore Health Naturally	4.95	_____
_____	How To Eat Your Way Back To Vibrant Health	4.95	_____
_____	How To Keep Healthy & Happy By Fasting	4.95	_____
_____	The Getting Back To Nature Diet	4.95	_____
_____	**How To Win Over IMPOTENCE/FRIGIDITY**	6.95	_____
	(Expanded Version with Full Color Section)		≡≡≡

Total for Books _____
Shipping & Handling _____
Total Enclosed $ _____

(We do NOT invoice. Check must accompany order, please.)

When using Credit Card, show number in space below.

☐ Check enclosed

☐ Master Charge

☐ VISA

When Using MasterCard
Also Give Interbank
No. (Just above your
name on card)

	Card	Month	Year
Ex-pires			

*POSTAGE & HANDLING Use the easy chart to figure postage, shipping and handling charges. Send correct amount and avoid delay.

TOTAL FOR BOOKS	Up to 5.00	5.01-10.00	10.01-20.00	20.01-35.00	Over 35.00
DELIVERY CHARGE	1.50	2.00	2.50	2.95	NO CHARGE

FOR ADDITIONAL SAVINGS: Orders Over $35.00 Are Now Postage-Free!

SHIP TO _____
      Mr./Mrs./Miss   (Please PRINT)

Address _____

City _____ State _____ ZIP _____

SALEM KIRBAN, Inc./Kent Road, Huntingdon Valley, Pennsylvania 19006